Fundamentals and Applications of Nanostructures Perovskite Oxide

THE AUTHORS

Dr. S. Thirumalairajan, presently working in the most prestigious DBT-Ramalingaswami Re-entry Faculty Fellow scheme (RRF), India in the Centre for Agricultural Nanotechnology (CANT), Coimbatore. He received a doctoral degree in Nanoscience and Technology - Physics (Interdisciplinary) from Bharathiar University, India. He is an expert in the fields of shape-dependent functional nanostructures thin film and powders for chemicals, biosensors, food Safety, and agricultural science applications. He has received two externally funded projects with a total cost of ~ Rs. 2 Crores to address issues for food safety applications. He has visited and worked as a senior postdoctoral fellow, and guest scientist in various countries like Brazil, USA, France, South Korea, Indonesia, and Taiwan, for the past nine years and the outcome of novel research resulted in India patent (4), articles in high impact international journals (34, citation ~1350, h-Index:24), national journals (2), Research book (3), book chapters (4), conferences oral/poster presentation (25), and received 10 research awards and distinctions for his credentials. He is a member of various scientific bodies in the national and international community.

Dr. K. Girija is an Assistant Professor of Physics, at Dr. N.G.P. Arts and Science College, Coimbatore. She completed doctor's degree in Physics - Nanoscience and Technology from the Department of Nanoscience and Technology, Bharathiar University, India. Her research also addresses the fabrication and evaluation of nanomaterials for use in multifunctional applications. Dr. K. Girija has received recognition as a Post-Doctoral Fellow from FAPSEP, Brazil, Senior Research Fellow, CSIR, Government of India and Junior Research Fellowship from DRDO - BU CLS, Coimbatore. She has published 24 articles, 2 patents, 3 books, 8 book chapters, and guides post graduate and Ph.D. scholars. She is a member of the Materials Research Society, Fellow of Asian Research Association, Electron Microscope Society and Indian Society for Technical Education.

Fundamentals and Applications of Nanostructures Perovskite Oxide

by

Dr. S. Thirumalairajan

Dr. K. Girija

2023

Scholars World®

A Division of

Astral International Pvt. Ltd.

New Delhi – 110 002

Cataloging in Publication Data--DK
Courtesy: D.K. Agencies (P) Ltd. <docinfo@dkagencies.com>

Thirumalairajan, S., author.
Fundamentals and applications of nanostructures perovskite oxide / by Dr. S. Thirumalairajan, Dr. K. Girija.
pages cm
Includes bibliographical references and index.
ISBN 9789359191300

1. Perovskite (Mineral) 2. Nanostructures. I. Girija, K., author. II. Title.

LCC TA455.P46T45 2023 | DDC 620.14 23

Published by : **Scholars World**
A Division of
Astral International Pvt. Ltd.
– ISO 9001:2015 Certified Company
4736/23, Ansari Road, Darya Ganj
New Delhi-110 002
Ph. 011-43549197, 23278134
E-mail: info@astralint.com
Website: www.astralint.com

Preface

Earlier the researcher believed the material properties can be altered only by varying the chemical composition. But later it was found that the material properties can be tuned by varying the size of the material without changing the chemical composition. The transition from micron-bulk materials to nanoscale lead to a number of changes in their physical properties. The major change is the increase in the surface area to volume ratio, as the size of the particle decreases to a regime where quantum confinement effects are predominant. There is a significant interest in identifying and manipulating perovskite oxide nanomaterials that have unique properties for applications in the field of nanoscience and technology. From both the scientific and technological point of view, it is worthwhile to learn the ways to control the size, morphology and improve functional properties. The authors discuss in this book, the preparation of different morphologies of nanostructured perovskite oxide $LaFeO_3$ powders and study its morphology-dependent physical properties and applications.

The book consists of six chapters and the contents of each chapter are as follows: the first chapter gives an overview of nanomaterials and perovskite oxide, the importance of morphology-dependent properties, and its application. A brief review of perovskite $LaFeO_3$ nanostructures with a focus on the chemical and physical method of synthesis, properties, and its applications has been presented. The second chapter describes in detail the preparation techniques for the different nanostructures and characterization tools employed. The third chapter explains the preparation and characterization of different morphologies of $LaFeO_3$ nanostructures such as nanocubes, nanorods, nanospheres, dendrites, flower-like nanostructures, microspheres and nanospheres prepared by the facile and environment-friendly wet chemical method. Also, a systematic investigation of growth mechanism, structural, compositional, and thermal analysis has been discussed. The fourth chapter deals with the physical properties such as surface area, optical, magnetic, and electrical properties of different morphologies of $LaFeO_3$ nanostructures. It is worthy to note that the prepared morphologies of $LaFeO_3$ nanostructures exhibit higher photocatalytic activity than Degussa P25 for the degradation of organic pollutant RhB under visible light irradiation and the $LaFeO_3$ nanostructures modified electrode-based dopamine biosensors have good sensitivity, selectivity, stability and low detection limit as discussed in the fifth chapter. Finally, sixth chapter gives the summary and conclusion of all the investigated results.

Dr. S. Thirumalairajan

Dr. K. Girija

Contents

Acknowledgement *v*

Preface *vii*

Acronyms and Abbreviations *xiii*

1. Nanomaterials and Perovskite Oxide: An Overview **1**

1.1 Introduction to Nanoscience and Technology

1.2 Importance of Controlled Size and Shape of Nanomaterials

1.3 Nanostructured Perovskite Pxide

1.4 Crystal Structure, Magnetic, Electrical and Optical Properties of $LaFeO_3$

1.5 Review of Literature

1.5.1 Chemical Synthesis of $LaFeO_3$ Nanostructure

1.5.2 Physical Preparation of $LaFeO_3$ Nanostructures

1.5.3 Photocatalytic Degradation of $LaFeO_3$ Nanostructures

1.5.4 Biosensing Nehaviour of $LaFeO_3$ Nanostructures

1.6 Motivation for this Book

1.7 Overview of this Book

References

2. Experimental and Characterization Techniques **17**

2.1 Introduction

2.2 Chemical Preparation Technique - Hydrothermal Process

2.2.1 Introduction

2.2.2 Hydrothermal Processing Parameters

2.2.3 Hydrothermal Apparatus

2.2.4 Advantages of the Hydrothermal Synthesis

2.3 Characterization Techniques

2.3.1 Structural Analysis - XRD, Raman and FTIR Spectroscopy

2.3.2 Compositional Analysis

2.3.3 Thermal Analysis

2.3.4 Morphological Analysis

2.3.5 Specific Surface Area Analysis

2.3.6 Optical Properties

2.3.7 Magnetic Properties

2.3.8 Electrical Properties

2.4 Conclusion

References

3. Controlled Synthesis and Growth Condition of Different Morphologies of $LaFeO_3$ Nanostructures **29**

3.1 Introduction

3.2 Experimental Procedure

3.2.1 Materials

3.2.2 Morphological Dependent Synthesis of $LaFeO_3$ Nanostructures

3.2.3 Characterization and Property Measurements

3.3 Results and Discussion

3.3.1 Morphological Analysis of different $LaFeO_3$ Nanostructures

3.3.2 Structural Analysis (XRD, Raman and FT-IR) of different $LaFeO_3$ Nanostructures

3.3.3 Compositional Analysis of different $LaFeO_3$ Nanostructures

3.3.4 Thermal Analysis of different $LaFeO_3$ Nanostructures

3.3.5 Effect of Surfactant Concentration and Growth Parameters of different $LaFeO_3$ Nanostructures

3.6 Conclusion

References

4. Morphology Dependent Physical Properties of $LaFeO_3$ Nanostructures **67**

4.1 Introduction

4.2 Surface Properties of $LaFeO_3$ Nanostructures

4.3 Optical Properties of different $LaFeO_3$ Nanostructures

4.4 Magnetic Properties of $LaFeO_3$ Nanostructures

4.4.1 Magnetic Materials

4.4.2 M-H Hysterias Loop

4.4.3 Shape of Hysteresis Loop

4.4.4 Type of Magnetic Domain

4.4.5 Magnetic Properties of different Morphologies $LaFeO_3$ Nanostructures

4.5 Electrical Properties of different $LaFeO_3$ Nanostructures

4.5.1 Introduction

4.5.2 Experimental Procedure

4.5.3 Results and Discussion

4.5.4 Dielectric Behaviour

4.5 Conclusion

References

5. Morphology Dependent $LaFeO_3$ Nanostructures for Photocatalytic and Biosensing Application **99**

PART- I

5.1 Morphology Dependent Photodegradation Study of $LaFeO_3$ Nanostructures for Rhodamine B under Visible Light

5.1.1 Introduction

5.1.2 Fundamentals of Photocatalysis

5.1.3 Steps Involved in the Heterogeneous Catalysis Process

5.1.4 Perovskite Oxide Photocatalytic Materials

5.1.5 Reaction Mechanism of Photocatalytic Reactive RhB

5.1.6 $LaFeO_3$ as Visible Light Photocatalyst

5.1.7 Measurement of Photocatalytic Activity

5.1.8 Results and Discussion

5.1.9 Conclusion

PART- II

5.2 Electrocatalytic Performance of different Morphologies of $LaFeO_3$ Nanostructures

5.2.1 Introduction

5.2.2 Modification of Electrode Surfaces with Nanomaterials

5.2.3 Electroanalytical Techniques

5.2.4 Cyclic Voltammetry (CV) and Linear Sweep Voltammetry (LSV)

5.2.5 Amperometric

5.2.6 $LaFeO_3$ Electrocatalysts for Dopamine Biosensing

5.2.7 Experimental Procedure

5.2.8 Results and Discussion

5.2.9 Conclusion

References

6. Summary and Conclusion **129**

Index **133**

Chapter 1

Nanomaterials and Perovskite Oxide: An Overview

1.1 Introduction to Nanoscience and Technology

Nanoscience and technology can be defined as the direct control of materials and devices on a molecular and atomic scale. It also includes fabrication of functional nanostructures with enhanced physical and chemical properties, synthesis and processing of nanoparticles, supra molecules chemistry, self assembly and replication techniques. The sintering of nanostructured metallic alloys, use of quantum effects, creation of chemical and biological templates and sensors, surface modification and nanostructured films are also the field of interest under nanotechnology. The materials and devices produced at nanoscale level take advantage of physical, chemical and biological principles whose causes are found in the nanometer scale.

Nanomaterials could be defined as the materials with at least one of its dimensions in the range of a nanometer (10^{-9} m). Thus the material need not be so small that it cannot be seen, it can be a large surface or a long wire whose thickness is in the scale of a nanometer.

- ✰ Materials that are nanoscale in one dimension include nano wires and nanotubes.
- ✰ Materials that are nanoscale in two dimensions are layers, such as thin films or surface coatings.
- ✰ Materials that are nanoscale in three dimensions are particles for example precipitates, colloids and quantum dots (small particles of semiconductor materials).

The properties of nanomaterials are very much different from those at a larger scale. The two principal factors that cause the properties of nanomaterials differ significantly from its counter parts.

- ☆ Surface effect (Surface to volume ratio)
- ☆ Quantum effect (Quantum confinement)

These factors can change or enhance properties such as reactivity, strength and electrical characteristics.

Surface Area to Volume Ratio

Nanomaterials have a relatively larger surface area when compared to the same volume or mass of a material produced in a bulk form. (Surface effect: causing smooth properties scaling due to the fraction of atoms at the surface).

Quantum Confinement

In nanocrystals, the electronic energy levels are not continuous as in the bulk. But are discrete (finite density of states) because of the confinement of electronic wave function to the physical dimensions of the particles. This phenomenon is called Quantum confinement and therefore nanocrystals are also referred as quantum dots (QDs).

Properties of the Materials can be different at Nanoscale

Nanostructured materials have properties that are different from those of bulk materials. Most nanostructured materials are crystalline in nature and they posses unique properties.

- ☆ Crystal structure of nanoparticles is same as bulk structure with different lattice parameters.
- ☆ The inter atomic spacing decreases with size due to long range electrostatic forces and the short-range core-core repulsion.
- ☆ The melting point of nanoparticles decreases with size.
- ☆ The electronic structure of nanoparticles is dependent on its size and the ability of nanocluster to react, depends on cluster size.
- ☆ The large surface area to volume ratio, the variations in geometry and the electronic structure of nanoparticles have a strong effect on catalytic properties.
- ☆ The electronic structure of nanomaterials is different from its bulk form.
- ☆ The density of energy states in the conduction band changes.
- ☆ Nanoclusters of different sizes will have different electronic structures and different energy level separations.
- ☆ The ionization potential at nanoscale is higher compared to the bulk materials.

- ✰ The magnetic moment of nanoparticles is found to be very less when compared with its bulk size.

1.2 Importance of Controlled Size and Shape of Nanomaterials

Nanomaterials are regarded as one of the most promising new materials in the 21st century. A significant reason is that it has large surface area along with the interfacial state. Hence, electrical, optical and magnetic properties are different from their bulk counter parts. It is well known that, the size and shape controlled synthesis is gaining incredible interest in nanostructured materials research. The size and shape dependent properties of the nanocrystals can be tuned by changing the preparatory conditions [1]. The materials at the nanoscale exhibit low coordination number and polarisable surface contribution to their chemical potential. For example, the high surface area is of particular importance regarding heterogeneous catalytic reactions, because of the increase in interaction of reactive molecules and active sites on the catalyst surface [2]. Furthermore, the particle size not only affects their surface area, but also gives rise to new properties due to the quantum-size effect [3]. Considerable efforts have recently been devoted to the preparation of metal oxide, perovskite oxide and mixed oxide nanomaterials due to their both unique properties and technological applications [4]. Perovskite oxides including the transition metals and rare earths display a wide variety of complex structures and interesting electronic and magnetic properties associated with the changes in electronic structure and bonding in the presence of ordered defect complexes or extended defects. The nanostructured mixed oxides can generate new synergetic properties and improve the overall application performance that is not available from single metal oxide species, due to the appropriate combination of individual oxide components [5]. Furthermore, the size and shape dependent properties of nanomaterials rising expectations for a better performance generally are a consequence of quantum confinement within the particle.

The precise controlled synthesis of size, shape, chemical composition, crystal structure, and surface chemistry of nanomaterials gives their unique functional properties, which have become one of the most challenging issues faced by nanomaterials researchers. The properties of nanocrystals are drastically altered by changing in its shape and size. These nanocrystals are ideal candidates for many potential applications, like catalysis, energy storage, optoelectronics, sensing, magnetic resonance imaging and biomedicine [6].

A general picture in Figure 1.1 schematically illustrates the features and new phenomena of the perovskite oxide nanoscale materials. The size and shape dependent properties of perovskite oxide structures at the nanoscale, makes great demand in the area of synthetic methodology. Therefore, it is a great challenge to develop a "synthetic chemistry" for nanocrystals that is as precise to make building units. This allows the scientists to study the effect of synthesis parameters which imparts to the unique collective properties of monodisperse nanocrystals. There are two different approaches to synthesize nanocrystals: the "top-down" approach, which utilizes physical methods, and the "bottom-up" approach involving chemical methods.

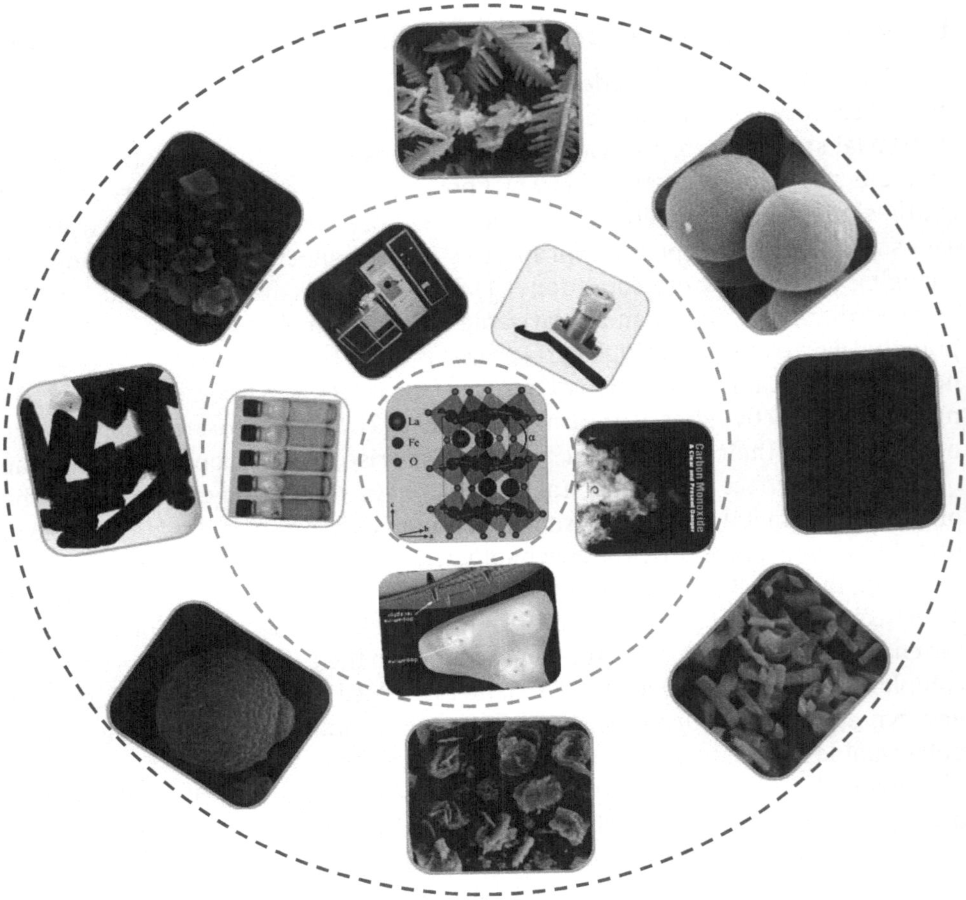

Figure 1.1. Representative Shapes of Perovskite Oxide Developed. These kinds of nanocrystals are utilized as key building blocks for catalytic, gas and bio-sensor applications.

In addition, the surfactant-assisted synthetic methods provide convenient and powerful pathway for the reproducible controlled synthesis of nanocrystals. These methods allows the metal oxide and perovskite oxide nanocrystals to be precisely adjusted in terms of their size, shape, composition, and phase structure at the nanometer scale. In general, the nanocrystals obtained by the surfactant-assisted route exhibits excellent crystallinity and monodispersity. In the following section, we discuss the development of facile and environmental friendly surfactant assisted routes for the shape and size controlled synthesis of perovskite oxide with different nanostructures. Critical reaction parameters that have strong effect on the growth of perovskite oxide nanocrystals includes precursor and surfactant concentration, the molar ratio of precursor to surfactant, and the reaction temperature and time. These are precisely adjusted to control their size and shape during the crystal nucleation-growth stages. In addition, to understand the formation process of nanocrystals in the bulk solution, we discuss the possible mechanism involved during the control of shape and size of the nanocrystals obtained from our surfactant-assisted approach.

1.3 Nanostructured Perovskite Oxide

The perovskite oxide (ABO_3) is named after a Russian mineralogist, Count Lev Aleksevich Perovski, and was discovered and named by Gustav Rose in 1839 from samples found in the Ural Mountains. Since then considerable attention has been paid to the perovskite family of compositions. The perovskite structure has the general stoichiometry ABO_3, where "A" and "B" are cations and "O" is an anion. The "A" and "B" cations can have a variety of charges and in the original perovskite mineral ($CaTiO_3$), the cation A is divalent or rare earths (La^{3+}, Ce^{3+}, Nd^{3+}, Sm^{3+} *etc.*) and the cation B is tetravalent or transitional metals (Fe^{3+}, Ni^{3+}, Al^{3+}, Cr^{3+}). However, in the present study, both A and B cations were adopted as trivalent state, and the cations 'A' were restricted to being rare earths. Recently, considerable efforts have been taken for the controlled synthesis of perovskite ABO_3 nanostructures and also investigation on the correlation between the materials properties and their shapes. It plays a vital role in many areas of chemistry, physics and materials science. The metal elements with different oxidation states can combine with oxygen to form a large diversity of oxide compound. These elements can adopt different structural geometrics with an electronic structure that can exhibit unique metallic, semiconductor or insulator character. In nanoscale, perovskite oxide exhibit unique physical and chemical properties due to their limited size and a high density of corner or edge surface sites. The particle size is expected to influence three important groups of properties in any materials.

Among the different forms of perovskite oxide such as $BiFeO_3$, $LaNiO_3$, $LaTiO_3$, $CaTiO_3$ *etc.*, Lanthanum orthoferrite, $LaFeO_3$ is a potential and a promising perovskite type oxide material. It displays significant physical and chemical properties making them of great importance to be used in advanced technologies such as solid oxide fuel cells, catalysts, chemical sensors, photocatalysts and biosensors. The morphology, size, porosity and specific surface area of $LaFeO_3$ materials are strongly influenced by the synthesis process. However, difficulties in controlling the nucleation and growth process of nanostructures exist. The construction of well-defined morphologies is still a great challenge in perovskite type nanomaterials.

1.4 Crystal Structure, Magnetic, Electrical and Optical Properties of $LaFeO_3$

Lanthanum orthoferrite, $LaFeO_3$ is one of the most common orthorhombic perovskite type oxide and a promising material with abundance of functionalities having a general formula ABO_3, which presents the large cation A in dodecahedral co-ordination and the small cation B in octahedral co-ordination. The transition metal ion B has several oxidation states that can be reduced and reoxidized. A stoichiometric ABO_3 compound with the combination of A^{n+} and B^{m+}, are given by n+m = 6, such as $LaFeO_3$, $LaCoO_3$, $CaTiO_3$, *etc.* The possible combination of A and B ions are controlled by the requirement that in the oxide framework the radii of A and B ions should be more than 0.90 Å and 0.51 Å respectively [7]. The traditional view of perovskite lattice is that it consists of small cations B within oxygen octahedra, and larger cations A which are XII fold coordinated by oxygen. This structural family is named after the $LaFeO_3$ which exhibits an orthorhombic structure with space group Pnma [8].

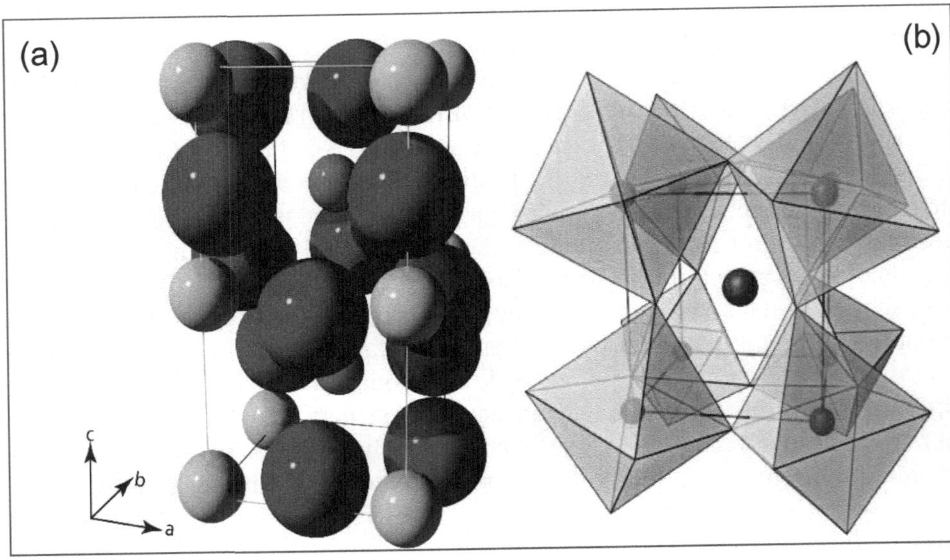

Figure 1.2. Crystal Structure of $LaFeO_3$ (a) Unit cell of orthorhombic perovskite showing the blue spheres represent the A cations, yellow spheres represent the B cations, with red spheres representing oxygen ions (b) Perovskite distortion FeO6 octahedra, the lanthanum atoms occupy the cavities between the octahedral and the projection along *b*-axis: Courtesy: in. wikipedia.com.

The crystal structure of $LaFeO_3$ can be described as consisting of orthorhombic structure along [121] direction. These oxides adopt a distorted perovskite-type structure with the space group *Pnma.* In which the rare-earth atom occupies interstitial sites between layers of corner-sharing FeO_6 octahedra. The arrangement of cation produces pairs of FeO_6 octahedra [9]. Each octahedron shares edges with three neighbouring octahedral in the sample plane and one face with an octahedron in an adjacent plane as shown in Figure 1.2 (a). Face sharing occurs along the b-axis (Figure 1.2 b). The La atom in this structure is surrounded by 12 O atoms, with eight of these atoms forming an inner coordination shell while the remaining four create an outer coordination shell. Based on structural arguments, it has been proposed that only the eight inner coordination shell (O atoms), which adopt distorted square anti-prismatic coordination geometry, interact with the La atom [10]. Because the rare-earth orthoferrites adopt an orthorhombic perovskite structure, the cation sites are highly substitutable and allowing the materials to be tuned for specific applications [11]. In order to rationally choose appropriate substitution schemes, it is necessary to understand how the electronic and bonding structures of these materials change with the identity of the rare-earth atom.

Magnetic Properties

When a material is placed within a magnetic field, the magnetic forces of the electrons will be affected. This effect is known as Faraday's Law of Magnetic Induction. However, materials can react quite differently in the presence of an external magnetic field. This reaction is dependent on a number of factors, such as the atomic and molecular structure of the material and the net magnetic field associated with the atoms. The magnetic moments associated with atoms have three origins. These are the electron motion, the change in motion caused by an external magnetic field, and the spin of the electrons. In most atoms the electrons occur in pairs. Electrons in a pair spin in opposite directions. So, when electrons are paired together, their opposite spins cause their magnetic fields to cancel each other. Therefore, no net magnetic field exists. Alternately, materials with some unpaired electrons will have a net magnetic field and will react more to an external field. Most materials come under the five kinds of magnetism classified as follows and as shown in Figure 1.3.

- ☆ **Diamagnetic:** Characterised by atoms having no inherent magnetic moments.
- ☆ **Paramagnetic:** Characterized by non-interacting atomic moments.
- ☆ **Ferromagnetic, antifeeromagnetic and ferrimagnetic:** Characterized by interacting magnetic moments.

Diamagnetic materials have a weak, negative susceptibility to magnetic fields. Diamagnetic materials are slightly repelled by a magnetic field and the material does not retain the magnetic properties when the external field is removed. In diamagnetic materials all the electrons are paired so there is no permanent net magnetic moment per atom. Diamagnetic properties arise from the realignment of electron paths under the influence of an external magnetic field. Most elements in the periodic table, including copper, silver, and gold, are diamagnetic.

Paramagnetic materials have a small, positive susceptibility to magnetic fields. These materials are slightly attracted by a magnetic field and the material does not retain the magnetic properties when the external field is removed. Paramagnetic properties are due to the presence of some unpaired electrons, and also from the realignment of the electron paths caused by the external magnetic field. Paramagnetic materials include magnesium, molybdenum, lithium, and tantalum.

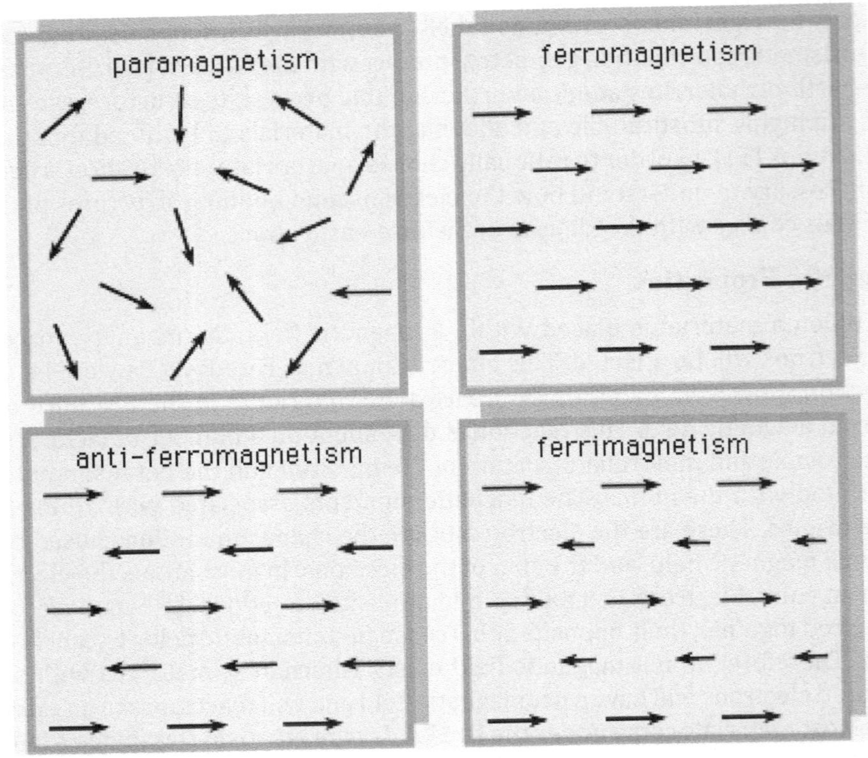

Figure 1.3. Schematic Representation of Spin Orientation in different Types of Magnetism.

Ferromagnetic materials have a large, positive susceptibility to an external magnetic field. They exhibit a strong attraction to magnetic field and are able to retain their magnetic properties after the external field has been removed. Ferromagnetic materials have some unpaired electrons so their atoms have a net magnetic moment. They get their strong magnetic properties due to the presence of magnetic domains. In these domains, large numbers of atom's moments (10^{12} to 10^{15}) are aligned parallel so that the magnetic force within the domain is strong. When a ferromagnetic material is in the unmagnitized state, the domains are nearly randomly organized and the net magnetic field for the part as a whole is zero. When a magnetizing force is applied, the domains become aligned to produce a strong magnetic field within the part. Iron, nickel, and cobalt are examples of ferromagnetic materials. Components with these materials are commonly inspected using the magnetic particle method.

Depending on the details of the crystal structure (which determines the size and sign of the exchange integral), exchange energy is minimum when electronic spins are aligned parallel or anti-parallel. This type of magnetic ordering is called **ferrimagnetism**. The magnetic structure is composed of two magnetic sublattice (called A and B) separated by oxygen. The exchange interactions are mediated by the oxygen anions. When this happens the interaction is called indirect super exchange interaction. In ferrimagnets, the magnetic moments of the A and B sublattices are not equal and results in a net magnetic moment.

Ferrimagnetism is therefore similar to ferromagnetism. Magnetite is a well known ferrimagnetic material with unequal sublattice corresponding to Fe^{3+} and Fe^{2+} site. If the A and B sublattice moments are exactly equal but opposite, the net moment is zero. This type of magnetic ordering is called antiferromagnetism, which occurs in materials like $LaFeO_3$ perovskite.

$LaFeO_3$ is a member of rare-earth orthoferrites ($RFeO_3$) which is a family of canted antiferromagnets with unusual magnetic properties and their magnetic behavior have been extensively studied in the 1950s and 1960s due to the scientific interests [12]. Recently, $RFeO_3$ have attracted much attention for their exchange bias, multiferroic and magneto-optical [13] properties which are very important for the development of new magnetic memory devices, low-power consumption spintronics devices and magneto-optical sensors. The magnetic structure in orthoferrites can be conventionally described by two interpenetrating pseudo-cubic face centred sub-lattices in which each Fe^{3+} ion is surrounded by six O^{2-} ions. This results in a collinear arrangement of the two sub-lattices, giving an antiferromagnetic ordering. However, the FeO_6 octahedra can be tilted to different degrees depending on the size of the rare earth cation, leading to a net magnetic moment [14-16].

Electrical Properties

The electronic structure of simple perovskite-type oxides near the fermi level is formed mainly by mixing of frontier *d*-orbitals of the transition metal and 2*p*-orbitals of oxygen. In $LaFeO_3$ perovskite oxide, iron is one of the first row transition element. The electronic configuration of Fe^{3+} ions is $1s^2\ 2s^2\ sp^6\ 3s^2\ 3p^6\ 3d^5$ and that of Fe^{2+} ions is $1s^2\ 2s^2\ sp^6\ 3s^2\ 3p^6\ 3d^6$. Thus, for the iron oxide, it is the Fe 3d electron that determines the electronic and magnetic properties. In an isolated atom, these 5d orbitals are degenerate, *i.e.*, their energies are equal. In oxide form, the electron of Fe ions interact with those of the electron of oxygen ions and this leads to a change in the energy state of transition metal. The cations (Fe ions) are octahedrally coordinated to negatively charged O^{2-} ions. Because of the electrostatic field arising due to this coordination, the Fe d orbitals no longer have the same energy. The d electron in orbitals directed along the coordinate axes experiences a greater repulsion from the negatively charged oxygen ions. This determines the electrical properties of $LaFeO_3$ materials.

Optical Properties

One of the most fascinating and useful aspects of nanomaterials is their optical properties. Applications based on optical properties of nanomaterials include optical detectors, sensors, imaging, phosphors, solar cell, photocatalysis, photoelectrochemistry display and biomedicine. The optical properties of perovskite nanomaterials depend on parameters such as feature size, shape, surface characteristics, and other variables including doping and interaction with the surrounding environment or other nanostructures. Likewise, shape causes have dramatic influence on the optical properties of metals, perovskite mixed metal nanostructures. Figure1.4 exemplifies the difference in the optical properties of metal and semiconductor nanoparticles. In the CdSe semiconducting nanoparticles, a simple change in size alters the optical properties of the nanoparticles. When metal nanoparticles are enlarged, their optical properties change only slightly as observed for the different samples of gold nanospheres in Figure 1.4. However, when anisotropy is added to these nanoparticles, such as growth of nanorods, the optical properties of nanoparticles change dramatically.

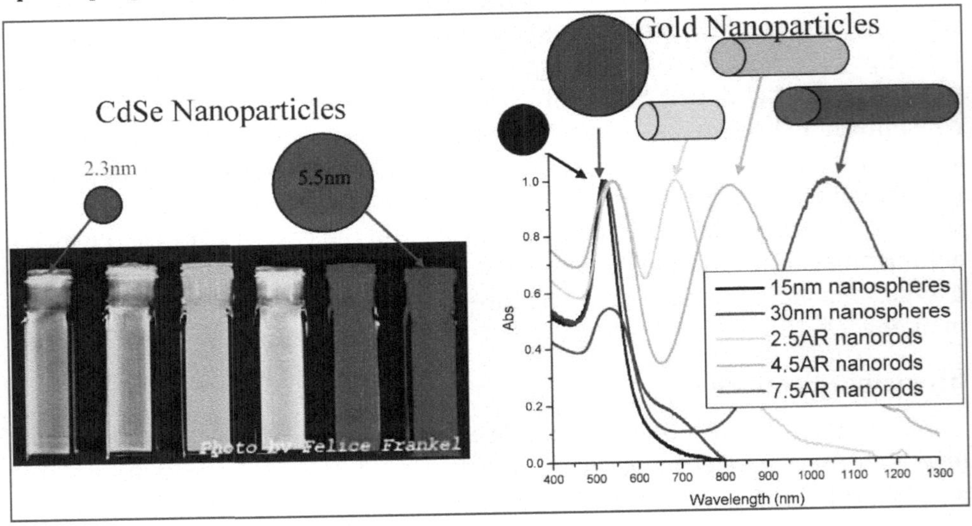

Figure 1.4. Fluorescence Emission of CdSe Quantum Dots of Various Sizes and Absorption Spectra of Various Sizes and Shapes of Gold Nanoparticles (Chem. Soc. Rev., 2006, 35, 209–217).

1.5 Review of Literature

An extensive survey on the literatures pertaining to the synthesis, properties and applications of $LaFeO_3$ nanostructures have been carried out and summarized as follows

1.5.1 Chemical Synthesis of $LaFeO_3$ Nanostructure

Over the past decades, many chemical methods such as sol-gel, co-precipitation, combined polymerization, pyrolysis of metallo-organic precursors, sonochemical method, microemulsion method and hydrothermal routes [19-23] have been developed to prepare $LaFeO_3$ nanostructures with controlled physical and chemical processes. These techniques uses metal oxides as starting precursors and requires several annealing processes at high temperatures for long periods of time with frequent intermediate grinding. These processes have several problems, *e.g.*, poor homogeneity and high porosity of the samples, no control on the particle size, *etc.* The development of innovative processing methods through chemistry permits to lower the preparation temperature and to improve homogeneity and reproducibility of the ceramic products, with the synthesis of ultrafine and chemically pure powders of mixed-metal oxides at low temperatures. Among the various conventional techniques, the hydrothermal technique is an aqueous-based precipitation route allowing control over the nucleation, growth and aging process. Surfactants as soft templates are involved in the formation of different morphologies by self-assembly process. The combination of surfactant-assisted self assembly under hydrothermal condition has been used by many researchers to synthesize different morphologies of other materials [24-27]. But in the present work citric acid, urea, PVA and CTAB were used as surfactant. In the synthesis of different $LaFeO_3$ nanostructures, as it is an inexpensive organic stabilizer, needs no complicated procedure, reactivity at relatively low temperature. There is no literature currently reporting the effect of surfactants on the formation of different morphologies of $LaFeO_3$ nanostructures.

1.5.2 Physical Preparation of $LaFeO_3$ Nanostructures

Physical techniques involved in the preparation of perovskite oxide thin films are pulsed laser deposition, chemical vapour deposition, thermal evaporation, e-beam evaporation, molecular beam epitaxy and sputtering deposition technique *etc.*, as reported in the literature [28-33]. However, the sputtering deposition technique requires several heating and grinding steps to ensure the homogeneous mixing of the various oxides [29]. Also, all the physical techniques require a very high calcination temperature and a long soaking to obtain the desired final products with good crystalline structure. Processes involving a higher calcination temperature and a longer soaking have several problems, *e.g.*, poor homogeneity and adhesion of the samples, no control of the particle size, *etc.*, which indirectly affects the functional properties of the obtained thin films. To lower the preparation temperature and to obtain ultrafine and chemically pure thin films of $LaFeO_3$, much effort has been spent recently on the development of physical methods. For this reason, it is of great significance to synthesize high quality $LaFeO_3$ thin films for potential applications. Enrico Traversa *et al.*, [33] fabricated $LaFeO_3$ thin films by RF magnetron sputtering method on Al_2O_3 substrates. Two distinct phases were noticed in the NO_2 response curve of $LaFeO_3$ thin films attributed to surface process of NO_2 and oxidation process of the bulk. They report that the annealing temperature and time of $LaFeO_3$ sputtered thin films determine the NO_2 sensitivity. In the present work we prepared different morphologies of $LaFeO_3$ nanostructure thinfilms by RF sputtering technique and investigate the morphology dependent properties and its potential in CO gas sensing applications.

1.5.3 Photocatalytic Degradation of $LaFeO_3$ Nanostructures

TiO_2 is considered to be an ideal semiconductor photocatalyst due to its photo stability, environment friendliness, low cost, and strong oxidizing ability [34]. However, there are two main bottleneck factors hindering its practical application, including high photoinduced charge recombination rate and low efficiency for utilizing solar light. Numerous studies have been performed to handle these problems including optimization of synthetic conditions, particle size and different morphology [35] coupling other semiconductors with narrow energy band gap and so on. In addition, it is regrettable that TiO_2 can exhibit high photocatalytic activity only under ultraviolet (UV) light irradiation. It is well known that $LaFeO_3$ perovskite oxide samples exhibit high photocatalytic activity towards degradation of organic compounds under visible light irradiation and can work in various environments steadily as reported by many researchers [36-39]. The semiconductor materials such as TiO_2 and ZnSe have been demonstrated to be an active photocatalyst [40, 41]. However, TiO_2 and ZnSe pertaining to faster charge carrier recombination exhibits activation only under ultraviolet irradiation due to their large band gap and shows reusability in aqueous solution. Therefore, choosing a material such as $LaFeO_3$ perovskite oxide with appropriate crystal structure, morphology, particle size and bandgap energy will dramatically improve its photocatalytic activity. However, to the best of our knowledge morphology different $LaFeO_3$ nanostructures towards visible light photocatalytic activities has not been explored in open literature.

1.5.4 Biosensing Behaviour of $LaFeO_3$ Nanostructures

Electrochemical biosensors constructed with inorganic nanomaterials have attracted extensive attention because they have not only excellent electrochemical and electrocatalytic properties [42]. Recently, numerous nanomaterials have been used to fabricate electrochemical biosensors because nanomaterials possess high surface-volume ratio and unique physical and chemical properties. It is well known that the electrochemical performance of a nanomaterial-based electrochemical biosensor is dependent on the component, structure, size, and shape of the nanomaterials [43-45]. At present, third generation electrochemical biosensors have been constructed using metals, semiconductors, and magnetic nanomaterials [43]. However, it is rare to employ inorganic perovskite nanomaterials for the construction of electrochemical biosensors although some inorganic nanomaterials have also been used to fabricate biosensors [46]. Interestingly, inorganic perovskite nanomaterials form layered structures more easily. Moreover, the biosensors based on layered porous nanostructures compared with those constructed with the common shape (*e.g.*, sphere, rod, and wire) nanomaterials were usually have better electrochemical performance and wider applications. Because the layered porous matrix has a higher surface area and more chemical active sites, and permits access to the bound neurotransmitter compounds. Recently perovskite-type oxide materials, such as (La, Sr) MnO_3, (La, Sr) CoO_3, (La, Sr) FeO_3 (LSF) and their solid solutions, are known electrode materials for solid oxide fuel cells (SOFC's) [47-48].

However, little attention has been paid to its application in electrochemical sensor. Based on its application in the electrode materials for fuel cells, we suppose the application of perovskite-type oxide materials may be as the electrode modifier and it is an interesting job to explore its application from this aspect. It was reported the synthesis of $LaFeO_3$ nanoparticles of ~22 nm in size characterized by XRD and TEM [48]. A novel glassy carbon electrode modified with $LaFeO_3$ nanoparticles was constructed and characterized by electrochemical impedance spectroscopy and cyclic voltammetry. The modified electrode exhibited strong promoting effect and high stability towards electrochemical oxidation of dopamine (DA) which gave reversible redox peaks with a formal potential of 0.145 V (Vs Ag/AgCl) in pH 7.0 PBS in the linear range of 1.5×10^{-7} to 8.0×10^{-4} M and the detection limit was 30×10^{-8} M. the interference by ascorbic acid was eliminated efficiently. The method was used to determine DA in dopamine hydrochloric injection and showed excellent sensitivity and recovery. In our present work, we reported the morphology dependent dopamine biosensing application for $LaFeO_3$.

1.6 Motivation for this Book

The perovskite oxides at nanoscale level have been the focus of study by researchers due to its functional properties and their properties are strongly related to the shape and size. A variety of interesting physical and chemical properties have been witnessed on varying the external stimuli such as temperature and pressure. The properties have not only opened the door for various applications but also for the better understanding of the material at nanoscale level. Some of the interesting properties include morphology, magnetic, electrical, optical, catalytic, gas sensing and biosensing.

1.7 Overview of this Book

The aim of present book was to achieve different morphologies with controlled size and improved physical properties of nanostructured perovskite oxide $LaFeO_3$ powders and thin films using hydrothermal and RF sputtering processes to enhance dopamine bio-sensing and Rhodamine B photocatalytic activity. We believe that the present work has provided an overview on various aspects of nanoscience and technology of $LaFeO_3$ perovskite oxide, starting from the controlled synthesis, analysis of the morphology-properties relationship, and use of nanomaterials in important applications relying on the novelty of nanostructures.

References

[1] Y. Masuda, K. Kato, K. *Cryst. Growth Des.* **8** (2008) 3213-3218.

[2] T. Nakanishi, Y. Masuda, K. Koumoto, K. *Chem. Mater.* **16** (2004) 3484-3488.

[3] V. Mohanrao, S. H. Han, M. P. Habib, *Cryt. Eng. Comm,* **14** (2012) 86-89.

[4] S.Thirumalairajan, K. Girija, I. Ganesh, D. Mangalaraj, C. Viswanathan, A. Balamurugan, N. Ponpandian, *Chem. Eng. J.* **209** (2012) 420-428.

[5] Y. Masuda, K. Kato, *Cryst. Growth Des.* **20** (2008) 1057-1063.

[6] S. O. Cho, E. J. Lee, H. Lee, J. G. Kim, Y. Kim, *J. Adv. Mater.* **18** (2006) 60-65.

[7] Q. Xia, M. Zhiying, F. Yuxin, D. Zhang, J. Ma, *et al.*, *Langmuir* **28** (2012) 5218-5226.

[8] Y. Wang, J. Zhu, L. Zhang, X. Yang, X. Wang, *Mater. Lett.* **60** (2006) 1767-1770.

[9] P. Sharma, J. Yeo, D. M. Kim, C. H. Cho, *J. Mater. Chem.* **22** (2012) 2838-2843.

[10] R. J.H. Voorhoeve, *Advanced materials in catalysis. Acade.* press. New York. 1977.

[11] L. G. Tejuca, J. Less. *Common Met.* **146** (1988) 261.

[12] Y. Masuda and K. Kato, *Chem. Mater.,* **20** (2008) 1057-1063.

[13] D. Treves, *J. Appl. Phys.* **36** (1965) 1033-1039.

[14] Y.S. Didosyan, H. Hauser, H. Wolfmayr, J. Nicolics, *Sens. Actuators A* **106** (2003) 168.

[15] S. Mathur, M. Veith, R. Rapalavicuite, H. Shen, G.F. Goya, W.L.M. Filho, T.S. Berquo, *Chem. Mater.* **16** (2004) 1906-1913.

[16] S. Mathur, H. Shen, N. Lecerf, A. Kjekshus, G.F. Goya, *Adv. Mater.,* **14** (2002) 1405.

[17] I. Bakonyi, E. Toth-Kadar, T. Tarnoczi, L. Varga, A. Cziraki, I. Gerocs and B. Fogarassy, *Nanostruct. Mater.* **3** (1993) 155.

[18] M. Manivel Raja, K. Chattopadhyay, B. Majumdar and A. Narayanasamy, *J. Alloys Comp.* **297** (2000) 199 -204.

[19] N. N. Toan, S. Saukko, V. Lantto *Physica B* **327** (2003) 279-282.

[20] J. Yang, H. Zhong, M. Li, L Zhang, Y. Zhang React Kinet *Catal Lett* **97** (2009) 269-274.

[21] C. Ting, Z. Zhou, Y Wang *et al.*, *Sens. and Actuators B* **143** (2009) 124-131.

[22] S. Bai, S. Bingjie, P. Yang, L. Zhiyong, D. Li, C. Afin., *Sci. China Ser B: Chem* **52** (2009) 2106-2113.

[23] Li, X.; Tang, C.; Ai, M.; Dong, L.; Xu, Z. *Chem. Mater.* **22** (2010) 4879-4889.

[24] A. Hossein, S. Hadi, P. Kameli, A. Poddar, M. Acet. *et al.*, J. *Phy. D: D Appl. Phys.* **43** (2010) 245002 (5pp).

[25] W. Hu, Y. Chen, H. Yuan, G. Zhang, G. Li *et al.*, *J. Solid state. Chem.,* **83** (2010) 1582.

[26] J. Ding, X. Li, H Shu, J Xie, H. Zhang *Mater. Sci. and Engineer.* B **171** (2010) 31-34.

[27] P. Song, Q. Wang, Z. Yang, *et al.*, *Physcia E* **41** (2009) 1479-1483.

[28] Z.X. Wei, Y.Q. Xu, H. Y. Liu *et al., J. Hazardous Mater.* **165** (2009) 1056-1061.

[29] M. Sivakumar, A. Gedanken, W. Zhong, *et al., J. Mater. Chem.*, **14** (2004) 764-769.

[30] X. Chu, S. Zhou, W. Zhang, H. Shui., *Mater. Sci. and Engineer. B* **164** (2009) 65-69.

[31] L. Sun, H. Qin, K. Wang, *et al., Mater. Chem. and Phy.* **125** (2011) 305-308.

[32] Fa-tang Li, Y. Liu, Z. Sun, R. Liu, D.S. Zhao, *et al., Mater. Lett.* **65** (2011) 406-408.

[33] E. Travera, S. Matsushima, G. Okada, *et al., Sens. And Actuators B* **24-25** (1995) 661.

[34] X. Yan, J. Chen, Y. Qi, J. Cheng, Z. Meng, *et al., J. European Ceramic Society* **30** (2010) 265-269.

[35] J. Chandradass, K. H. Kim, *Mater. Chem. and Phy.* **122** (2010) 329-332.

[36] K. M. Parida, K.H. Reddy, S. Martha, *et al.,* Inter. J. *Hydro. Energy* **35** (2010) 12161.

[37] G. Wang, J. Sun, W. Zhang, S. Jiao, B. Fang, *Microchim. Acta.* **164** (2009) 357-362.

[38] V. Hariharan, S. Radhakrishnan, C. Sekar *et al., Talanta* **85** (2011) 2166-2174.

[39] G. Deng, Y. Chen, M.Tao, C. Wu, X. Shen, H. Yang, M. Liu, *Electro. Acta* **55** (2010) 1120-1124.

[40] D. Wang, Xiangfeng Chu and M. Gong *Nanotechnology* **17** (2006) 5501-505

[41] Z. Yang, Y. Huang, B. Dong, *et al., Mater. Res. Bull.* **41** (2006) 274-281

[42] A. Chainani, M. Mathew, D. D. Sarma *Phy. Rew. B* **48** (1993) 20.

[43] D. Wang, M. Gong, *J. Appl. Phy.,* **109**, (2011) 114304.

[44] Z. Li, Y. Liu,P. Liu, W. Chen, S. Feng, *et al., RSC Advances,* **2** (2012) 3447–3450.

[45] S. Carter, A. Selcuk, R.J Charter, J. Kajada, *et al., Solid State Ionics* **59** (1992) 753.

[46] J. E Elshof, M. H. Lankhorts, H. J. Mouwmeester *Solid State Ionics* **99** (1997) 15.

[47] C. R. Raj, T. Okajima, T. Ohsaka, J Electroanal Chem **543** (2003) 127.

[48] G. Wang, J. Sun, W. Zhang, S. Jiao, B. Fang, *Microchim. Acta.* **164** (2009) 357-362.

[49] H. Obayashi, Y. Sakurai, T. Gejo, *J. Solid State Chem.* **17** (1976) 299.

[50] T. Arakawa, S. Tsuchi-ya, J. Shiokawa, *Mater. Res. Bull.* **16** (1981) 97.

[51] G. Martinelli, M. C. Carotta, M. Ferroni, Y. Sadaoka, *Sens. Actuators B* **55** (1999) 99.

[52] H. Aono, E. Traversa, M. Sakamoto, Y. Sadaoka, *Sens. Actuators B* **94** (2003) 132.

[53] X. Niu, W. Du, W. Du, *Sens. Actuators B* **99** (2004) 399.

[54] D. Fino, N. Russo, G. Saracco, V. Specchia, *J. Catal.* **217** (2003) 367.

[55] R.-F. Huang,W.-Y. Howng, *J. Mater. Res.* **11** (1996), 3077.

Chapter 2

Experimental and Characterization Techniques

2.1 Introduction

The shape and size controlled synthesis of functional nanostructures such as nanocubes, nanospheres, nanowires, dendrites, flowers and nanorods has gained importance due to its enhanced functional properties and it is a challenging task in modern nanomaterials process [1]. The importance of morphology controlled synthesis was realised from the fact - the properties of nanostructures and the subsequent applications strongly depend on the preparation methods. Hence, exploring successfully the synthesis method and tailoring the properties of nanostructured materials have been an active area of research by materials scientists and chemists. Recently, different physical and chemical methods are being identified and explored for the preparation of nanostructures of diverse materials [2].

The reaction of well-defined morphologies is still a great challenge in perovskite type nanomaterials. So far, nanostructures of various morphologies have been synthesized by a variety of physical and chemical routes. Among the various conventional techniques available, the hydrothermal approach [3] and RF sputtering technique [4] have attracted much more attention due to their distinct advantages such as effective control over the size and shape of particles, relatively low reaction temperature, high homogeneity and well-crystallized products with defined composition. Hence, in the present work, we have made adroit use of the above techniques for the synthesis of perovskite type oxide nanostructures with different morphologies. However, efficient handing of these methods in the preparation of different nanostructures requires a thorough understanding of the physics and chemistry involved in the processing. Therefore, the initial part of this chapter explains the comprehensive understanding of the physical and chemical parameters involved in the synthesis and fabrication of nanomaterials using hydrothermal and RF magnetron sputtering techniques. The role of processing parameters like temperature, precursor concentration and aging time has also been elaborated. In addition, the requirement instrument for both the techniques has been discussed.

The chemical and physical preparation methods have reached significant advances in terms of homogeneity in the structural and especially electronic and magnetic properties of the particles forming the solid as well as in the size distribution [5]. However, much work is still essential to fully control the synthesis variables and understand their influence on the functional properties of the final products. The problems related to perovskite oxide stoichiometry, the metal oxidation state or the presence of impurities and amorphous phases coexisting with the crystalline phase are often encountered during the synthesis. A glimpse into these problems would indicate that the structural and electronic properties as well as the primary particle size distribution are strongly dependent on the preparation methods. Final part of this chapter give an overview of the experimental techniques along with the characterization tools used to analyse the prepared samples.

2.2 Chemical Preparation Technique - Hydrothermal Process

2.2.1 Introduction

Over the past decade, many wet chemical methods such as sol-gel, co-precipitation, combined polymerization, pyrolysis of metallo-organic precursors, hydrothermal routes *etc.* [3] have been developed to prepare perovskite oxide nanostructures with controlled physical and chemical characteristics. Among the various conventional techniques available, the hydrothermal technique is an aqueous-based precipitation route allowing control over the nucleation, growth and aging process. It is typically carried out in a pressurised vessel called autoclave with the reactions in aqueous solution. The temperature in the autoclave can be raised above the boiling point of water, reaching the pressure of vapour saturation. Hydrothermal synthesis is widely used for the preparation of metal oxide nanoparticles which can easily be obtained through hydrothermal treatment. The technique has been proved useful to control the grain size, particle morphology, crystalline phase and surface chemistry through regulation of the solution composition, reaction temperature, pressure, solvent properties, additives and aging time. The hydrothermal process is a non conversional method to obtain inorganic nanocrystalline materials. The hydrothermal technique is an appropriate and one of the most important tools for advanced materials processing, particularly owing to its advantages in the processing of nanostructured materials for a wide variety of technological applications. The hydrothermal technique not only helps in processing monodispersed and highly homogeneous nanoparticles, but also acts as one of the most attractive technique for processing nano-hybrid and nanocomposite materials. The term 'hydrothermal' is purely of geological origin. Hydrothermal processing can be defined as any heterogeneous reaction in the presence of aqueous solvents or mineralizers under high pressure and temperature conditions to dissolve and recrystallize (recover) materials that are relatively insoluble under ordinary conditions. It is defined as any heterogeneous chemical reaction in the presence of a solvent (whether aqueous or non-aqueous) above the room temperature and the pressure greater than 1 atm in a closed system [6]. Among various technologies available today in advanced materials processing, the hydrothermal technique occupies a unique place owing to its advantages over other conventional techniques.

2.2.2 Hydrothermal Processing Parameters

The hydrothermal synthesis process involve the solubility in water of almost all inorganic substances at elevated temperature and pressure, and the subsequent crystallization of dissolved material in the fluid. Water at elevated temperature plays an essential role in the precursor material transformation. However, hydrothermal processing is rather complex, as greater number of interacting parameters play the essential roles.

Pressure is essential for solubility, the supersaturation range directing the crystallization process as well as for the thermodynamic stability of the product phase.

Temperature plays an important role in kinetics of product formation as well as on the thermodynamic stability of the product phase. Formation of temperature gradient within the hydrothermal chamber is a key factor in this process.

Precursor concentration is essential for the solubility, the supersaturation range directing the crystallization process as well as for the thermodynamatic stability of the product phase.

Time is also an important parameter because the synthesis of kinetically stable phase are favoured in short term processes while the thermodynamatically stable phases are generally formed in long term experiments corresponding to the chosen temperature-pressure regime. This is caused by re-dissolution and recrystallization of the already formed phases.

2.2.3 Hydrothermal Apparatus

The key limitation of conventional hydrothermal method has been the need for time consuming empirical trial and error methods, as a means for process development. Material processing under hydrothermal condition requires a vessel capable of containing a highly corrosive solvent at high temperature and pressure. Ideal hydrothermal apparatus popularly known as 'autoclave' should have the following characteristics:

1. Inertness to acids, bases and oxidizing agents.
2. Ease of assembly and disassembly.
3. Sufficient length to obtain a desired temperature gradient.
4. Leak-proof at the required temperature and pressure.
5. Rugged enough to bear high pressure and temperature experiments for long periods with no damage, so that no machining or treatment is needed after each experimental run.

In the hydrothermal synthesis, the precursor materials are taken in a container with solid to water proportion of about 1:10. The closed containers are placed into the sealed stainless steel autoclave and place into the oven. The most successful materials are corrosion resistant, high strength alloys such as 300 series stainless steel, iron, nickel, cobalt based super alloys, titanium and its alloys. In the majority of hydrothermal experiments, the mineralizers used are highly corrosive and it can attack the vessel, which is inimical for obtaining high purity materials. Therefore, a suitable lining for the inner wall of autoclave or separate liners are required for the autoclave. Some of the preferred liners are silver, copper, teflon, nickel, gold, pyrex, graphite, platinum *etc.*

In the present study, the hydrothermal apparatus used for the synthesis is a Teflon-lined stainless-steel autoclave of 40 ml capacity. The teflon liner, stainless steel casing and the accessories of hydrothermal apparatus used in the present work are shown in Figure 2.1. The precursor solution filled up to ¾th of the teflon container, sealed in the stainless-steel casing was kept in a conventional hot air oven at required temperature and time.

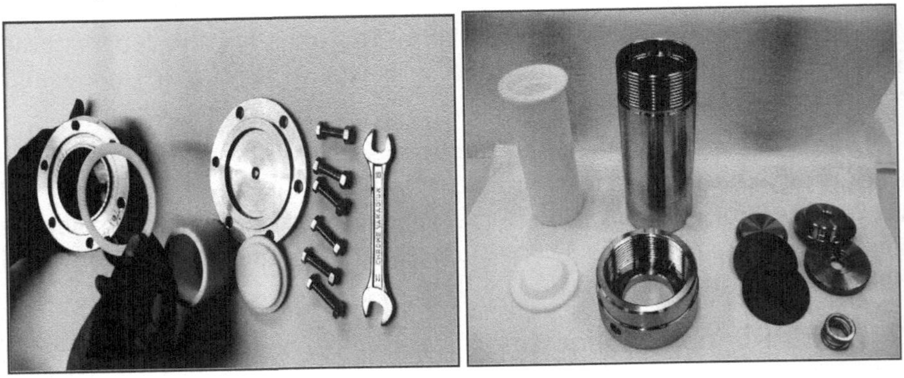

Figure 2.1. Photograph of Teflon Line, Stainless Steel Casing and the Accessories of the Hydrothermal Apparatus Used in the Present Study.

2.2.4 Advantages of the Hydrothermal Synthesis

The hydrothermal processing of advanced materials have plenty of advantages and can be used to achieve high purity product and homogeneity, crystal symmetry, meta stable compounds with unique properties, a lower sintering temperature, dense sintered powders, a wide range of chemical composition, single-step processes, sub-micron to nanoparticles with narrow particle size distribution using simple equipment, lower energy requirement, fast reaction time, lowest residence time, the growth of crystals with polymorphic modifications, as well as for the growth of crystals with low to ultra low solubility and a host of other applications. In general, hydrothermal technology in the 21st century has altogether offered a new perspective in fabrication of functional nanostructures. The behavior of the solvent under hydrothermal conditions deals with the aspects like structure at critical, supercritical and sub-critical conditions, dielectric constant, pH variation, viscosity, coefficient of expansion, density, *etc.*, is to be understood with respect to pressure and temperature. Similarly, the thermodynamic studies yield rich information on the behaviour of solutions with varying pressure and temperature conditions. Temperature fluctuations of the oven have negative consequences due to rise in the temperature leads to a higher dissolution rate disturbing the dynamic equilibrium of dissolution-crystallization process, while lowering of temperature leads to higher supersaturation.

2.3 Characterization Techniques

2.3.1 Structural Analysis - XRD, Raman and FTIR Spectroscopy

The phase, purity and crystallinity of the samples were structurally characterized using X-ray diffraction, Raman spectroscopyand Fourier transformed infrared spectroscopy.

(a) X-ray Diffraction (XRD)

X-ray diffraction is a well-known and most frequently applied technique for structural characterization as shown in Figure 2.2. This technique is a versatile, non-destructive analytical technique for identification and quantitative determination of the various crystalline forms (phase) of compounds present in powder and solid samples. Fine powdered samples were spread on a glass slide using double sided sticky tape and X-ray diffractograms were collected using a Bruker Germany D8 Advance X-ray diffractometer, with diffraction angle 2 θ in the range 20-80° using Cu Kα_1 radiation (λ = 1.5406 Å) and comparing the diffraction peaks with the database such as JCPDS (Joint committee on powder diffraction standards) or ICDD (International centre for diffraction Data). From the resulting diffractogram, the average crystallite size (*D*, nm) can also be estimated using the Scherrer's formula and the density of prepared crystallites were also evaluated [8].

(b) Fourier Transformed Infrared Spectroscopy (FT-IR)

The FT-IR spectrum was also recorded to monitor the structure confirmation of perovskite nanostructures as in Figure 2.3. Infrared radiation interacts with chemical bonds to cause stretches, bends and various other atomic vibrations. For a vibration to give rise to absorption of infrared radiation, it must cause a change

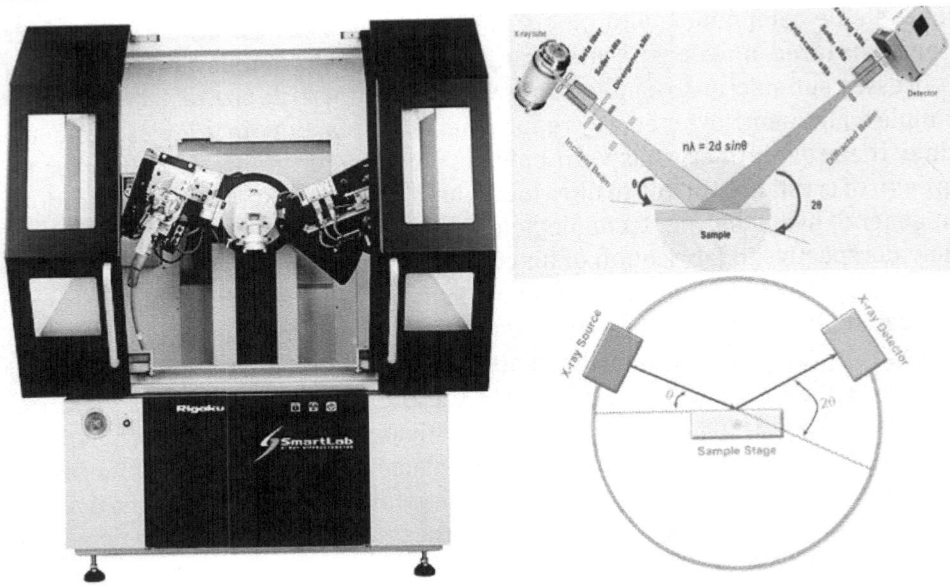

Figure 2.2: Photograph and Working Process of X-ray Diffraction (XRD), Courtesy: rigaku.ac.in.

in the dipole moment of the molecule. The larger this change the more intense the absorption band will be. In addition, the size of the peaks in the spectrum is a direct indication of the amount of the material present. The spectrum was recorded using a Thermo Nicolet 200 FTIR spectrometer. Samples were recorded as a KBr disc (1:10, sample/KBr), using 16 scans per sample at a resolution of 4 cm^{-1} over the range 4000-400 cm^{-1}.

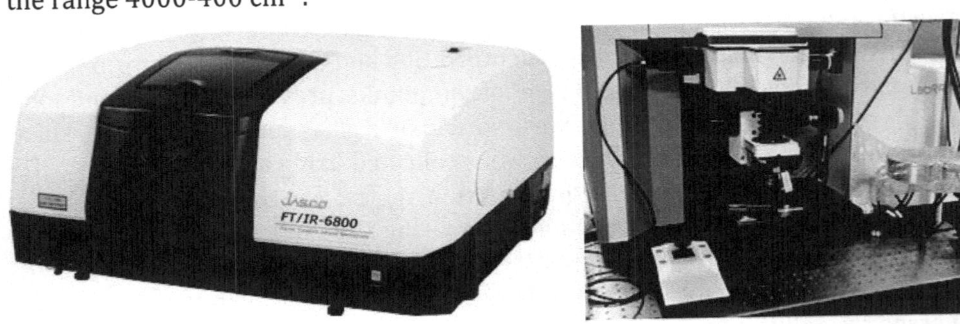

Figure 2.3. Photograph of FTIR and Raman Spectroscopy,

(c) Raman Spectroscopy

The Raman spectroscopy has the advantage of being very sensitive to structure distortion and oxygen motion. So, a detailed investigation of electron excitation is very important for the understanding of perovskite-type materials [9]. Raman spectroscopy, although similar to IR (both measure the vibrational energies of molecules) does not cause a change in the dipole moment of a molecule. For a vibration to be Raman active, the polarisability of the molecule must change with the vibrational mode. The technology involves the measurement of scattered light from a sample, which is illuminated with a monochromatic light source such as a laser. Raman spectroscopy is based on the Raman effect, which is the inelastic scattering of photon molecules [9]. Raman spectra were obtained for the prepared samples placed on a glass wafer using LabRAM HR 800 Raman microscope with 514 nm Ar ion laser (10 mW).Typically, a small amount of powdered sample was mounted on a glass slide using a spatula tip. Using the attached microscope and camera, the instrument was focused on the powder sample through the objective lens (x10). One scan was recorded with a detector exposure time of 1sec.

2.3.2 Compositional Analysis

X-ray Photoelectron Spectroscopy (XPS)

The elemental composition, purity and chemical status of perovskite oxide nanostructures were carried out using X-ray photoelectron spectroscopy on ESCA + Omicron UK XPS system with Mg Kα source and photon energy 1486.6 eV. All the binding energies were referenced to the C 1s peak at 284.6 eV of the surface adventitious carbon. Furthermore, because the binding energies differ not only from chemical species to species, but also with the bonding conditions in which the elements are found. This technique provides information on the actual compounds present on the surface. The analysis area was ~ 500 μm in diameter. For survey spectra, energy of 100 eV and a 0.4 eV step size was used. Charge compensation was achieved using a low energy electron flood gun. Quantitative surface chemical analyses were evaluated from the high resolution core level spectra, following the removal of a non-linear Shirley background. These characteristic peaks correspond to the electron configuration of the electrons with the atoms. The number of detected electrons in each of the characteristic peak is directly related to the amount of element within the area (volume) irradiated. The manufacturer's 'Avantage' software was used to plot the results. The software incorporates the appropriate sensitivity factors and corrects the electron energy analyser transmission function.

2.3.3 Thermal Analysis

Thermogravimetric and Differential Thermal Analysis (TG/DTA)

Most of the perovskite oxide nanostructured materials may get dehydrated to some degree without major alternations in their crystal structure. Thermal analysis comprise a group of techniques in which physcial properties of a substance is measured as a function of temperature, while the substance is subjected to a controlled tempeature programme. With thermal analysis, reaction kinetics, thermal stability, oxidative stability, presence of impurities and thermal expansive rates can be analyzed. In general, two types of thermal analysis are mainly used

- ✰ Thermogravimetric analysis (TGA)
- ✰ Differential thermal analysis (DTA)

TGA analysis measures the weight change in the material as a function of temperature under controlled atmosphere. A derivative weight loss curve obtained can be used to infer the point at which weight loss is most apparent. This study gives detail about the stability, chemical composition, degradation temperature and absorbed moisture content in materials. Similarly, DTA is a thermoanalytic technique, in which the material under study and an inert reference are made to undergo identical thermal cycles while recording the temperature difference between sample and reference. This differential temperature is then plotted against time or temperature (DTA curve or thermogram). The change in the sample, either exothermic and endothermic can be detected relative to the inert reference. Thus, a DTA curve provides data on the transfomations that have occurred, such as glass transitions, crystallization, melting and sublimation. Finally TG/DTA curve was obtained using simultaneous thermal analyzer (TG/DTA, SDT Q 600 V20) from RT to 1000 °C under nitrogen atmosphere.

2.3.4 Morphological Analysis

(a) Scanning Electron Microscopy (SEM)

The scanning electron microscope (SEM) is a type of electron microscope that uses electrons instead of light to form an image and can provide significant informations regarding the surface of materials including fractured surfaces, bond failures and physical defects [10]. In addition, backscattered electron imaging is useful in distinguishing one material from another, since the yield of the collected backscattered electron increases monotonically with the atomic number. Backscattering imaging can distinguish elements from their atomic number differences, with good contrast in the image. SEM uses a highly focused electron beam to strike and interact with the sample which is contained in a high vacuum environment to form a high resolution image. The images formed by the scanning electron microscope are from secondary electrons, backscattered electrons, characteristic x-ray, Auger electrons and others that are emitted by the sample. It has many advantages over traditional microscopes. In the present study, morphology of the samples was obtained using scanning electron microscopy (SEM- JEOL JSM-6380LV).

(b) Transition Electron Microscopy (TEM)

TEM is used to identify and analyse the atomic level structure of materials. A very thin slice of the materials to the tested is exposed to a beam of electrons. When the electrons interact with a consistent material structure, a constant fraction of electron is transmitted back from the sample to a detector. Once a structural imperfection is encountered, the fraction of transmitted electron changes, resulting in the change of contrast. Diffraction contrast is useful in identifying structures and crystallographic features. Phase contrast is used for high magnification imaging of atomic columns. Higher magnification visual observation were conducted using transmission electron microscopy (TEM, JEM 2100 F) - performed with an acceleration voltage of 200 kV by placing a small amount of sample powder on a holey-carbon film on a copper grid by ultrasonication. All samples were carefully prepared for better inspection of the atomic level structure.

(c) Atomic Force Microscopy (AFM)

Atomic force microscopy (AFM) is used to acquire high-resolution real-space topographical images of various sample surfaces. It does not require the sample to be electrically conductive unlike scanning electron microscopy and can be carried out on any type of sample surfaces. A vacuum environment is not essential, thus it is convenient to carry out AFM measurements compared to TEM. The AFM probes the surface of a sample with a sharp tip, a couple of μm long and less than 10 nm in diameter. The tip is located at the free end of a cantilever, 100 to 200 μm long. Forces between the tip and the sample surface cause the cantilever to bend or deflect. A detector measures the cantilever deflection as the sample is scanned. From the measured cantilever deflection, a map of the surface topography can be generated. The force most commonly associated with AFM is the inter-atomic Vander Waals force. In contact mode, the cantilever is held less than a few Å from the sample surface where the inter-atomic force between the cantilever and the sample is repulsive in nature. In non-contact mode, the cantilever is held about few Å above the surface where the inter-atomic force is an attractive one. In tapping mode of operation, the cantilever is oscillated closer to the sample than in non-contact mode. Part of the oscillation extends into the repulsive regime, so the tip intermittently touches the surface. Tapping mode is a technological improvement in AFM. Tapping mode overcomes problems associated with friction, adhesion, electrostatic forces and other difficulties. A large area, high precision AFM (Model: Asylum research MFP 3D, USA) was used to study the surface morphology of selective samples.

2.3.5 Specific Surface Area Analysis

BET and BJH Methods

The determination of specific surface area of the samples has become a part of nanostructure characterization and an important tool among nanomaterials scientist. The specific surface area and pore size distribution in the sample are determined by using gas adsorption-desorption isotherms by BET and BJH methods. The gas adsorption technique is performed by the addition of a known volume of gas (adsorbate), typically nitrogen, to a solid material in a sample vessel at cryogenic temperature. At this temperature, weak molecular attractive forces will cause the gas molecules to adsorb onto a solid material. An adsorbate is introduced in a series of controlled dose, while the pressure in the sample vessel is measured after each dose. There is a direct relationship between the pressure and the volume of gas in the sample vessel. Buy measuring the reduced pressure due to adsorption, the ideal gas law can then be used to evaluate the volume of gas adsorbed by the sample. The resulting relationship of, volume of gas adsorbed Vs relative pressure at constant temperature is known as an adsorption isotherm. In this analysis, the specific surface area and pore size distribution of the sample can be computed. The Brunauer-Emmett-Teller (BET) isotherms of the powders were analyzed by nitrogen adsorption-desorption in a Micromeritics ASAP 2020. The prepared samples were degassed at 180 °C prior to nitrogen adsorption measurements. The isotherms were used to determine pore size distribution by the Barret-Joyner-Halender (BJH) method.

2.3.6 Optical Properties

Ultraviolet-Diffusion Reflectance Spectroscopy (UV-DRS)

The diffuse reflectance spectrum was recorded using a UV-Vis spectrophotometer (UV-2550, Shimadzu, Japan) for the prepared samples pressed into a pellet of uniform thickness and placing the pellet into the reflection sphere where it reflects light directly from the source, which is further detected and recorded. From the resulting diffuse reflectance spectra, the band gap of the perovskite samples could be evaluated.

2.3.7 Magnetic Properties

Vibrating Sample Magnetometer (VSM)

The VSM can be used to measures the magnetic properties of prepared samples with an applied magnetic field. The M-H loop (hysteresis loop) can be used to derive the essential magnetic parameters, namely, saturation magnetization, remanent magnetization and coercivity. The magnetic characterization (diamagnetic, paramagnetic, ferromagnetic or antiferromagnetic) of any material is best described in terms of their hysteresis loop. In the present work, magnetic properties of various nanostructures were analyzed using vibrating sample magnetometer (VSM-EG and G, PARC) with an applied magnetic field of 20 kOe at room temperature.

2.3.8 Electrical Properties

Impedance Spectroscopy (IS)

Impedance spectroscopy is a flexible tool for simultaneous electrical and dielectric characterization of materials. This powerful technique has been widely used to characterize the dielectric behaviour of single crystal, polycrystalline and amorphous ceramic materials. The experimental data contains three variables, real, imaginary component of the electrical response and frequency together with four inters related electrical response formalism: impedance (Z^*), admittance (Y^*), relative permittivity (ε^*) and electrical modulus (M^*). Consequently, data can be presented in many formats; in practice, it is often found that different formats may highlight different electrical characteristics of a sample. It is the most commonly used experimental technique to analyze the dynamics of the ionic movement in solids. The general approach is to apply a monochromatic (single frequency) voltage to the sample and measure the corresponding response current as the frequency is scanned. In the present work, Solartron-SI1260 impedance/grain phase analyzer with frequency range 100 μHz to 10 MHz, fully automated with a personal computer was used to study the electrical transport properties of the samples.

2.4 Conclusion

In summary, the physics underlying in the physical and chemical methods were discussed in the first segment. The growth parameters of the synthesis processes that need to be manipulated have been listed out, to make use of and achieve different morphology of nanostructures. In the second half of the chapter, brief note on various characterization techniques involved, the instruments used and the methods adopted during the measurement have been discussed.

References

[1] Y. Masuda, K. Kato, *Cryst. Growth Des.* **8** (2008) 3213-3218.

[2] T. Nakanishi, Y. Masuda, K. Koumoto, *Chem. Mater.* **16** (2004) 3484-3488.

[3] K. Byarappa, M. Yoshimura, *Hand book of hydrothermal technology,* Noyes publications, William Andrew publishing, 2001.

[4] K. Wasa, S. Hayakawa, *Handbook of sputtering deposition and t technology: Principles, Technology and Applications*, 1992.

[5] A.S. Edelstein, R.C. Cammaratra, *Nanomaterials: Synthesis, Properties and Appl.*, 1998

[6] K. Byarappa, T. Ohachi, *Crystal growth technology*, William Andrew publishing, Springer, 2003

[8] C. Kittel, *Introduction of solid state physics* 7th edition, 2007.

[9] Walter. S. Struve, *Fundamental of molecular spectroscopy*, 1989.

[10] Ludwig Reimer, *Scanning Electron Microscopy Physics of Image formation and Microanalysis*, 1988.

Chapter 3

Controlled Synthesis and Growth Condition of Different Morphologies of $LaFeO_3$ Nanostructures

3.1 Introduction

Nanomaterials are regarded as one of the most promising new materials in the 21st century. One significant reason is that nanomaterials have large surface area along with interfacial state, hence electrical, optical and magnetic properties of nanomaterials are superior than bulk materials. It is well known that the size and shape-controlled synthesis is gaining tremendous interest in nanostructure materials research [1-3]. Recently, considerable efforts have been taken for the controlled synthesis of perovskite $LaFeO_3$ nanostructures and investigate their shapes dependent properties [2]. However, it remains a big challenge to freely operate the materials at micro/nano level using facile methods. The synthesis and analysis of complex functional structures with various morphologies will offer an opportunity to build nanodevices using bottom up approaches [4]. Hence, the shape or size-controlled synthesis of anisotropic perovskite $LaFeO_3$ nanostructures is key research area in nanoscience and nanotechnology. Now, the challenging is to control these parameters of perovskite $LaFeO_3$ nanostructures with accomplished synthetic approaches to obtain different morphologies. Several techniques have been used for the growth of nanostructured $LaFeO_3$ nanostructures *viz.*, sol-gel, hydrolysis, thermal evaporation and metal-organic chemical vapour deposition [5-10]. Among these methods hydrothermal synthesis is an ideal technique for synthesizing $LaFeO_3$ perovskite nanomaterials with high purity, controlled morphology, high crystallinity, low temperature growth and less complicated. Also, surfactants as soft templates are involved in the formation of different morphologies by self-assembly process. Urea, CTAB, PVP and citric acid are common and superior structure directing agents for metal ions to form different morphologies. These surfactants are very important in controlling the dissolution, nucleation and crystallization process under the action of temperature and pressure [11]. The combination of surfactant-assisted self-assembly under hydrothermal condition has been used by many researchers to synthesize different morphologies [12]

The present chapter aims to provide an overview about the controlled synthesis and growth condition of different morphologies of $LaFeO_3$ such as nanocubes, nanorods, nanospheres, dendrites, flower like nanostructures and microspheres composed of nanoparticles prepared by facile, environmental friendly hydrothermal process. These nanostructures have been systematically investigated for morphological, structural, thermal and compositional analysis. The formation mechanism of different morphologies of $LaFeO_3$ nanostructures was proposed and the role of surfactant and growth parameters in controlling the particle morphologies also elucidated.

3.2 Experimental Procedure

3.2.1 Materials

Lanthanum nitrate hexahydrate ($La(NO_3)_3.6H_2O$), ferric nitrate hydrate ($Fe(NO_3)_3.9H_2O$) or potassium ferric cyanide $K_3[Fe(CN)_6]$ as starting materials. Urea ($(NH_2)_2CO$) citric acid ($C_6H_8O_7.H_2O$), hexadecyltrimethyl ammonium bromide (CTAB) and polyvinylpyrrolidone (PVP, K30) $(C_6H_9NO)_n$ were used as surfactant or structure directing agents. All the chemical were analytical pure grade (99.99 per cent) and used without any further purification. Double deionized water was used as solvent during the synthesis.

3.2.2 Morphological Dependent Synthesis of $LaFeO_3$ Nanostructures

The morphologically dependent perovskite $LaFeO_3$ nanostructures were synthesized by hydrothermal process with the help of soft surfactants. Lanthanum nitrate hexahydrate ($La\ (NO_3)_3.6H_2O$), potassium ferric cyanide $K_3[Fe(CN)_6]$ or ferric nitrate hydrate ($Fe(NO_3)_3.9H_2O$) were used as starting materials, while urea ($(NH_2)_2CO$), citric acid ($C_6H_8O_7.H_2O$), hexadecyltrimethyl ammonium bromide (CTAB) and polyvinylpyrrolidone (PVP, K30) $(C_6H_9NO)_n$ were used as soft surfactants. Six different reactions were performed for the synthesis of $LaFeO_3$ nanostructures. In the first reaction, equal molar amount (1:1) of lanthanum nitrate hexahydrate and potassium ferric cyanide were dissolved in 30 ml double distilled water under magnetic stirring. A transparent red solution was obtained after thoroughly stirring the solution. In the successive reactions, in addition to the above solution urea, CTAB, PVP and citric acid were separately added as surfactant. The concentration of all surfactant was varied as (1:1:1), (1:1:2) and (1:1:3). In the last reaction starting material $K_3[Fe(CN)_6]$ was replaced with $(Fe(NO_3)_3.9H_2O)$ and critic acid was added as surfactant. In the typical experiment, required quantities of surfactants were first dissolved in distilled water to form a solution of desired molar concentration and then added drop wise to the metal ion solution. The precursor solution was stirred magnetically for 30 min and then transferred separately to a Teflon-lined stainless steel autoclave for hydrothermal treatment to maintain at 180 °C for 12 h in an oven. After the hydrothermal treatment, the autoclave was cooled naturally to room temperature. The precipitates settled at the bottom was collected and washed repeatedly with distilled water and ethanol to remove the impurities and excessive surfactant. The samples were finally dried at 100 °C followed by calcination at 800 °C for 2 h to obtain different morphologies of $LaFeO_3$ powder and it was subjected for further characterization.

3.2.3 Characterization and Property Measurements

In order to understand the growth of these nanostructures, the products obtained at various stages of the chemical reaction were systematically characterized by different techniques. The surface morphology, chemical composition, particle size and structure of the prepared nanostructures were characterized by using scanning electron microscopy (SEM), energy dispersive X-ray analysis (EDX) (SEM- JEOL JSM-6380LV), transmission electron microscopy (TEM, JEM 2100 F) with an acceleration voltage of 200 kV by placing the powder on a copper grid. X-ray diffraction (XRD), Bruker Germany D8 Advance) with $CuK_{\alpha 1}$ radiation (λ = 1.54 Å). X-ray photoelectron spectroscopy (XPS) measurements were done on ESCA + Omicron UK XPS system with MgK_{α} source and photon energy of 1486.6 eV. All the binding energies were referenced to the C 1s peak at 284.6 eV of the surface adventitious carbon. Raman spectrum for the prepared samples was obtained by placing sample on a glass wafer using LabRAM HR 800 micro Raman with 514 nm Ar ion laser (10 mW). Fourier transformed infrared (FTIR) spectroscopy was performed on a Thermo Nicolet 200 FTIR spectrometer using KBr wafer technique. The spectrum was collected in the mid-IR range from 400 to 4000 cm^{-1} with a resolution of 4 cm^{-1}. The thermal stability of the prepared samples was determined from the simultaneous thermal analysis (TG/DTA, SDT Q 600 V20) from RT to 1000 °C under nitrogen atmosphere.

3.3 Results and Discussion

3.3.1 Morphological Analysis of different $LaFeO_3$ Nanostructures

(a) Nanocubes, Nanorods and Nanospheres

In this section, the surface morphologies of the prepared $LaFeO_3$ nanostructures synthesized by three different hydrothermal reactions using lanthanum nitrate hexahydrate and potassium ferric cyanide as a staring materials with urea and citric acid as a surfactants. The obtained SEM micrographs are shown in Figures 3 a-d cube like $LaFeO_3$ morphology was obtained for the direct hydrothermal synthesis

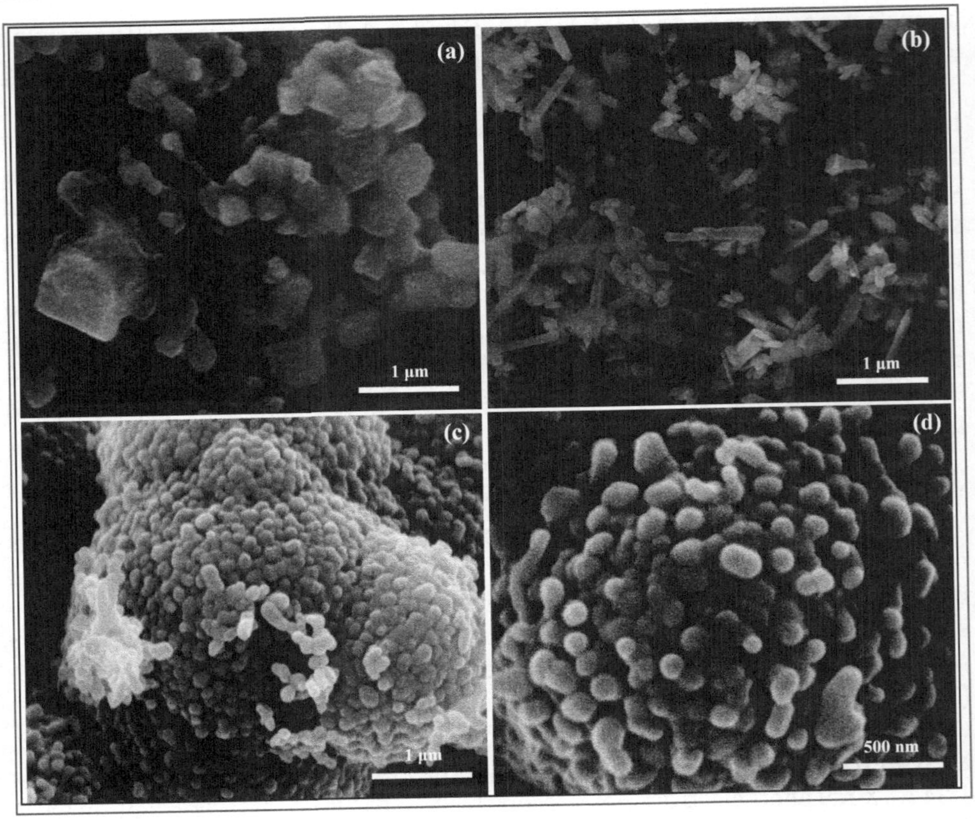

Figure 3.1. SEM Images of Three different $LaFeO_3$ Morphologies (a) Nanocubes, (b) Nanorods and (c and d) Low and High Magnification of Nanospheres.

of $LaFeO_3$ without using any structure directing agent and it has sharp edges and smooth surface as shown in Figure 3.1a. When urea was add as structure directing agent, the nanorods of $LaFeO_3$ were obtained as shown in Figure 3.1b. These nanorods have the aspect ratio (ratio between the diameter ~ 60 nm and length 240 nm) of ~ 4 as evidenced from the SEM image. The another structure directing agent citric acid yield the sphere shaped nanostructures with the average size of 52 nm as shown in Figures 3.1c and d. The above results confirmed the cube like morphology was obtained without using any structure directing agents. Further, urea and citric acid influences the change in morphology from cube to rod and sphere with orthorhombic structure as confirmed by XRD as discussed later in the structural analysis section. It is understood that surfactant ions originating from urea and citric acid play a major role in the formation of different morphologies. The formation of one dimensional of one dimensional nanorod was obtained. When urea was used as the structure directing agent. Urea ions play a major role in the formation of one-dimensional shape, *i.e.*, adsorbs to the surface of $LaFeO_3$ seed crystal and thereby decreases the surface energy in all directions expect one direction.

In the particular undisturbed direction, further growth takes place by means of agglomeration/attachment with other seed layers. The process continues to give one dimensional rod like structures. In another process, citric acid ions adsorbed on the surface of $LaFeO_3$ particle direct the self-assembly process of nanoparticles by molecular interaction to form nanospheres. These results confirm the structure directing agents are simple and effective way to control the nucleation and growth process [13]. The surfactants not only provide favourable sites for the growth of particle assemblies, but also influence the formation process, including nucleation, growth, coagulation and flocculation. All these different surfactants were found to have specific mechanism involved in the formation of nanostructures. During the synthesis process, the surfactants were adsorbed by the growing crystal and depending upon the precursor concentration and surfactant properties it can moderate the growth rate of crystal faces, which thereby helps in the size and shape control of nanostructures [13].

Further insights into the morphology and microstructure of the prepared nanostructures were provided by TEM and HRTEM images as shown in Figure 3.2. The morphology of $LaFeO_3$ resembles cube shaped nanostructure with primary nanounits ~ 90 nm and the secondary crystals with the average size of ~150 nm as shown in Figure 3.2a. The HRTEM image recorded at the tip of individual $LaFeO_3$ cubes is shown in Figure 3.2b, and the unidirectional alignment of lattice fringes confirms the polycrystallinity of the sample. The regular spacing of the observed lattice plane was 0.29 nm, which is corresponding to the lattice spacing of (110) plane of cubic $LaFeO_3$. The interface of primary $LaFeO_3$ nanounits involved in the formation of secondary $LaFeO_3$ nanocubes can be evidenced from the HRTEM image in the inset of Figure 3.2b. It gives additional information about the packing of nanounits. The trivial difference in contrast makes it possible to distinguish the edges and shape of the individual nanobuilding blocks. Each nanocrystal shows well-defined lattice fringes. The SAED pattern in the inset of Figure 3.2b demonstrates the growth axis is along (110) direction of $LaFeO_3$ cubes.

$LaFeO_3$ nanorod like structures as shown in Figure 3.2c was obtained, when urea was used as structure directing agent. A close observation reveals that single nanorods have diameter ~ 60 and length 240 nm. The alternate bright and dark region in the TEM and HRTEM image signifies the variation in the surface thickness of $LaFeO_3$ nanorods. The HRTEM image of the tip portion of nanorods shows the lattice planes with the interplanar spacing of 0.279 nm corresponding to the (121) direction.

The polycrystalline nature of nanorods was revealed by the discrete spots in the SAED pattern shown in the inset of Figure3.2d. The formation of nanospheres with average the particle size of 50 nm when citric acid was used as structure directing agent are shown in Figure 3.2e. The HRTEM image in Figure3.2 f shows the lattice fringes with a distance of 0.276 nm which matches with the (121) crystallographic plane of $LaFeO_3$. The corresponding SAED pattern in the inset of Figure 3.2f indicates

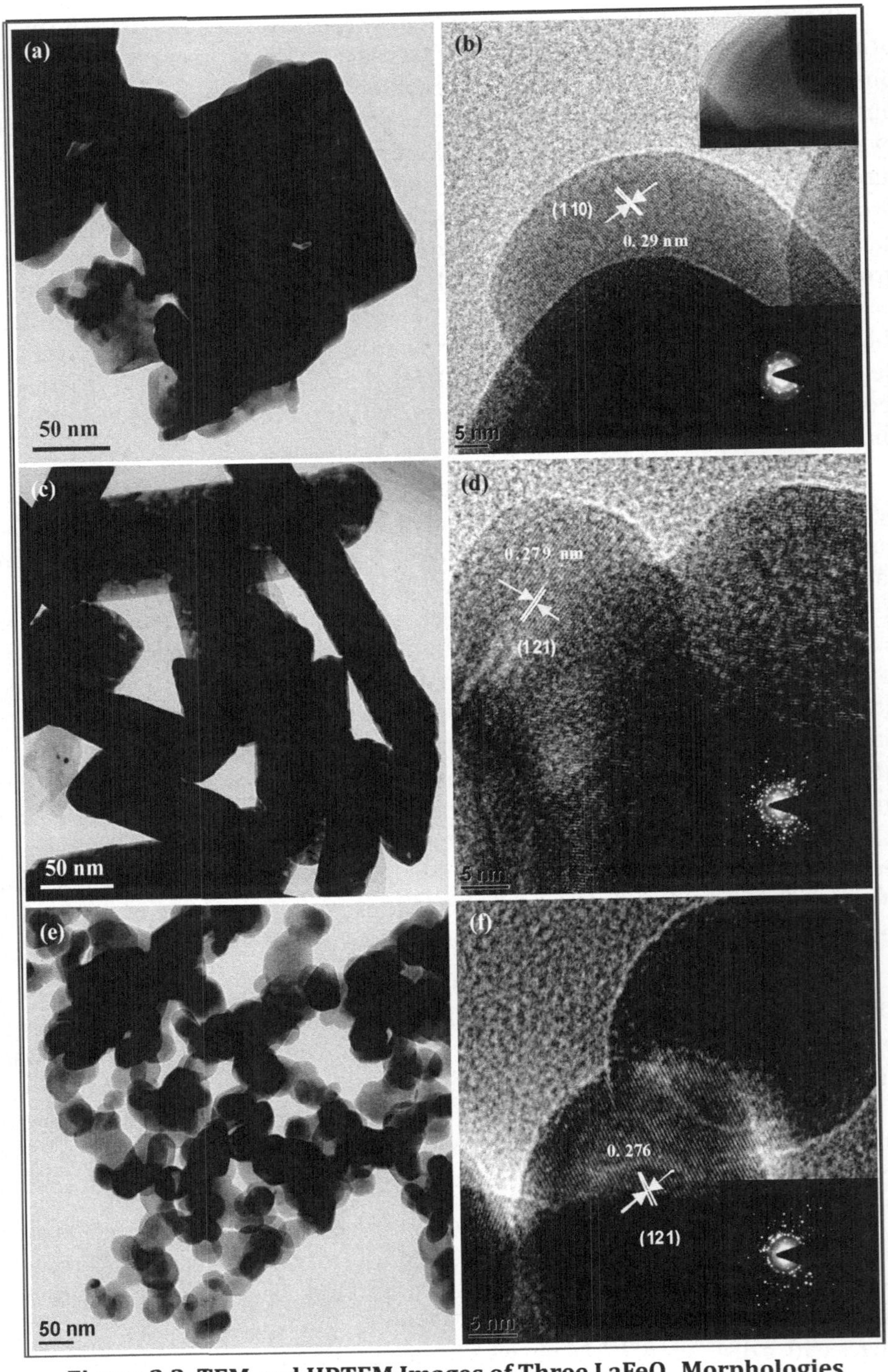

Figure 3.2. TEM and HRTEM Images of Three $LaFeO_3$ Morphologies (a-b) Nanocube, (c-d) Nanorod (e-f) Nanosphere and the Inset is Corresponding SAED Pattern.

the polycrystalline nature which may be due to the diffraction from nanospheres. Thus by correlating the outcome from TEM and HRTEM images it can be concluded that the observed results are in good agreement with SEM and XRD pattern.

(b) $LaFeO_3$ Dendritic Nanostructures

The another surface morphology of the $LaFeO_3$ nanostructures was prepared using lanthanum nitrate hexahydrate, potassium ferric cyanide as using staring materials and CTAB as surfactant. The observed SEM image for this sample is shown in Figure 3.3a, which clearly indicates the $LaFeO_3$ dendritic structures with long main trunk and short side branches. Further observation of high magnification image in Figure 3.3b reveals the clear, well defined and uniformly ordered dendrite structures with distinct trunk and branches distributed on both sides of the trunk. The overall length of the dendrite is about 3-4 μm, with an average trunk diameter 80 nm and the branches 1-1.5 μm in length and diameter of 85 nm. The branches became shorter and shorter from the bottom to the tip of the stem and no sub branches were formed on the main branch. The dendritic growth occurs at the

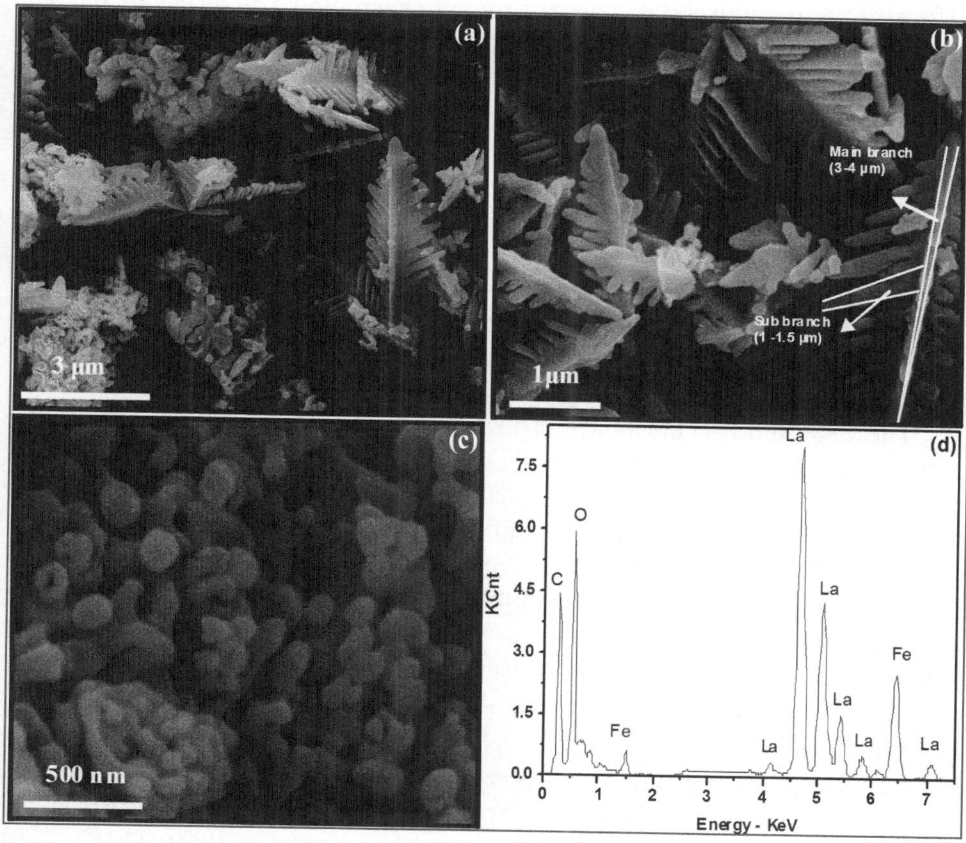

Figure 3.3. (a) SEM Micrograph for the $LaFeO_3$ Dendritic Nanostructures (a) Low Magnification, (b) High Magnification View of Dendrites Main Stem and Branch, (c) Aggregation of $LaFeO_3$ Nanoparticles, and (d) EDAX Pattern.

tips of the stem and branches as growth time increases. As the stem grows longer, new branches were continuously formed near the tip region. For the higher-order branches (leaves), the dimension can reach as small as a few nanometers. Amazingly, all the side branches are flanked by the trunk symmetrically with an angle of ca. 60°, implying that the $LaFeO_3$ dendritic crystals grow along a preferential (121) direction. The $LaFeO_3$ nanoparticles of size around 80-90 nm shown in Figure 3.3c formed under supersaturation condition ends up with dendritic structures due to interplay between diffusion kinetics and the oriented attachment of particles. The chemical composition of $LaFeO_3$ dendrites was examined by energy-dispersive X-ray spectroscopy. As seen in Figure 3.3d, the elements present in the samples are La, Fe and O with a molar ratio of 1:1:3 corresponding to the stoichiometric composition of $LaFeO_3$. The C peak in the spectrum can be attributed to the electric latex of the SEM sample holder. This is also in agreement with the result obtained from XPS analysis discussed in later section.

Further, the cationic surfactant CTAB ions absorbed on the surface of $LaFeO_3$ particle directs the self-assembly process of nanoparticles by molecular interaction to form dendrites with certain crystallographic orientation. The surfactant not only provides favourable sites for the growth of particle assemblies, but also influences the formation process, including nucleation, growth, coagulation and flocculation. The surfactant was found to have specific mechanism involved in the synthesis of nanostructures. During the synthesis process, the surfactant was absorbed by the growing crystal and depending upon the precursor concentration and surfactant properties it moderates the growth rate of crystal faces, which thereby helps in control the shape and size [13]. However, as the system is driven farther from equilibrium, dendritic growth can be promoted because the growth rate of crystals significantly exceeds the mass transport rate of ions that feed the growing crystals. In supersaturation-based crystallization, both the growth rate and the mass transport rate are governed by the concentration of the reactant [14].

The TEM and HRTEM image in Figure 3.4 provides further insight on the morphology of the prepared $LaFeO_3$ dendrites. The low-magnification TEM image in Figure 3.4 reveals the high yield of $LaFeO_3$ dendrites and the uniformity achieved through this approach. A higher magnification TEM image of a single dendrite is shown in Figure 3.4b, which clearly reveals the dendritic structure. The length of the $LaFeO_3$ dendritic trunk was ~ 3-4 μm. In Figure 3.4c, it is clear that the dendrite is highly symmetric and the angle between stem and branches was about 60°. The length and diameter of the branches are 1-1.5 μm and ~ 85 nm, respectively.

The HRTEM image recorded from the tip of the individual dendritic $LaFeO_3$ is shown in Figure 3.4d. The lattice fringes with the distance of 0.27 nm was observed and it corresponds to (121) crystallographic plane of $LaFeO_3$ dendrites. The corresponding SAED pattern in the inset of Figure 3.4d indicates the single crystalline nature due to the major diffraction spots corresponding to (121), (220) and (202). No diffraction spots were attributed to the secondary phase or impurity.

Thus by correlating the outcome from TEM and HRTEM images it can be concluded that the observed results are in good agreement with SEM results.

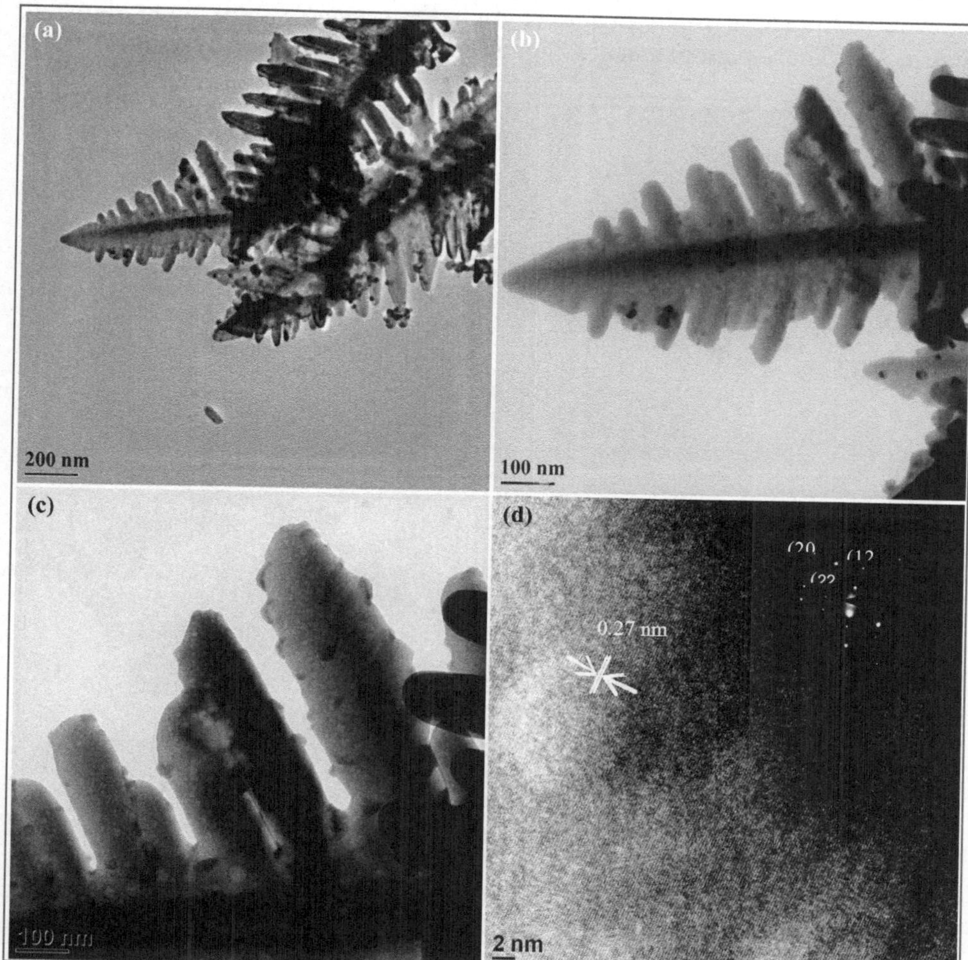

Figure 3.4. TEM Image of $LaFeO_3$ Dendritic Nanostructures with (a) Low Magnification, (b) High Magnification, (c) Branch, and (d) HRTEM Image of Dendrites showing Lattice Fringes (Inset) SAED Pattern.

(c) Flower like Morphology Comprised of $LaFeO_3$ Nanoplates

The flower like $LaFeO_3$ nanostructure was prepared by using lanthanum nitrate hexahydrate and potassium ferric cyanide as staring materials and PVP as surfactant. Figure 3.5 shows the corresponding SEM micrograph. The lower magnification image in Figure 3.5a indicates the morphology of $LaFeO_3$ samples were flower like structures with an average diameter of a micro-flower of 3 μm. A detailed view of individual flower like $LaFeO_3$ nanostructures can be observed from the high magnified SEM image as shown in Figure 3.5b. It clearly shows the $LaFeO_3$ flower like superstructures are comprised of two-dimensional thin nanoplates with an average thickness of 15 nm and length in the range of 70-80 nm as shown in Figure 3.5c. These nanoplates are aligned perpendicular to the spherical surface

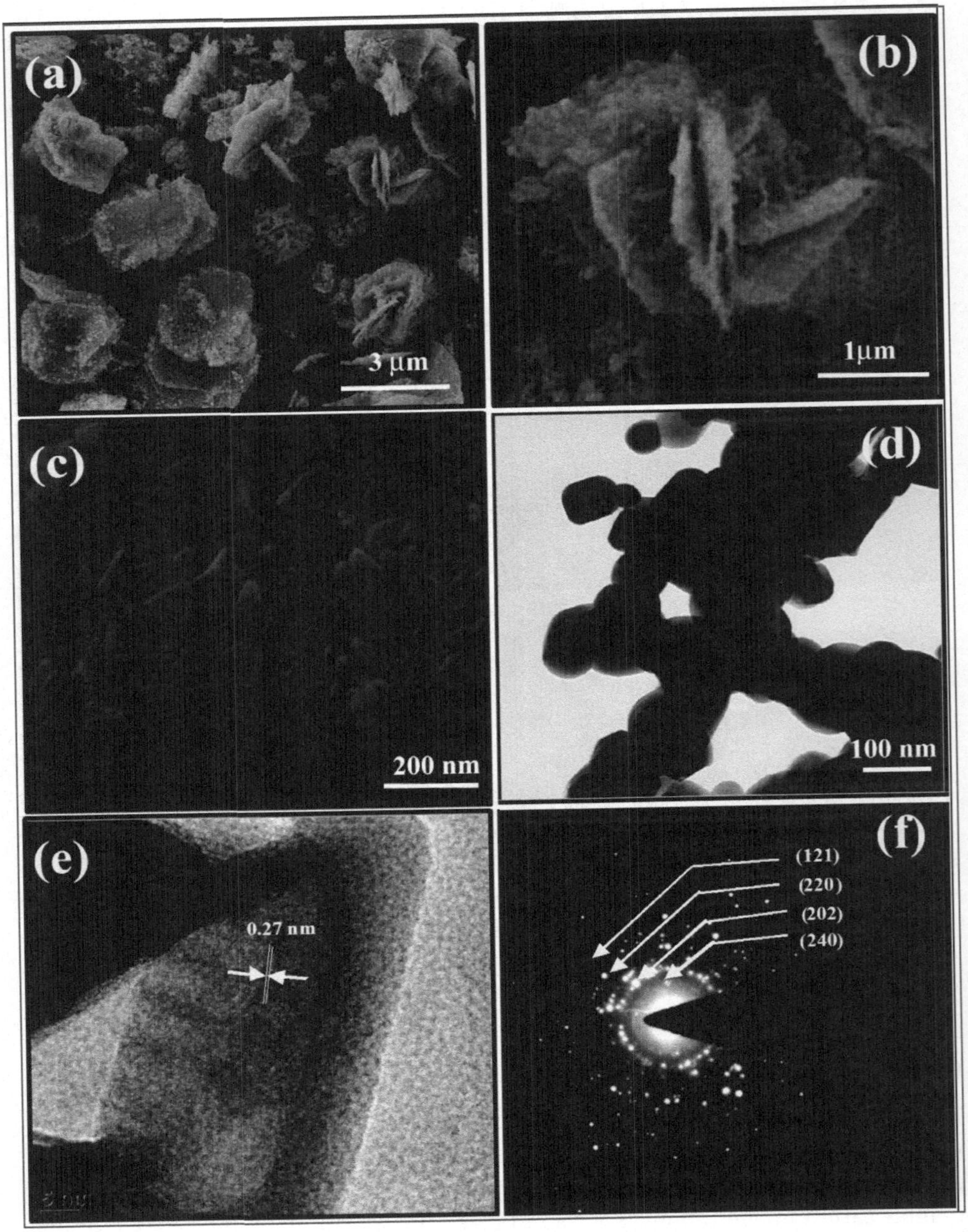

Figure 3.5. Flower like $LaFeO_3$ Nanostructures with (a) Low Magnification SEM, (b and c) High Magnification SEM, (d) TEM Image, (e) HRTEM and (f) SAED Pattern.

and hence involved in the formation of flower like structures via self-assembly of nanoplates. Significantly, these hierarchical nanoplates are sufficiently stable as they are not destroyed even after high temperature calcination (800 °C). The TEM image provided in Figure 3.5 d for further insight on detailed morphology and structure of flower like $LaFeO_3$ samples. It confirms that the entire structure was built from nanoplates with an average size of ~70 nm, also evaluated from XRD pattern as discussed in later section. These plates are loosely packed and connected to each other through the centre to form flower like structures. The corresponding high resolution TEM (HRTEM) image taken on the tip of a single plate shows that the regular spacing of the observed lattice plane was ca. 0.27 nm as shown Figure 3.5e, which was consistent with the (121) crystal plane of orthorhombic $LaFeO_3$ structure which could explain the broadening of XRD peaks. The diffraction rings are discontinuous and consists of rather sharp spots which indicate good crystallinity of $LaFeO_3$ nanoplates as observed from the selected area electron diffraction pattern (SAED) shown in Figure 3.5f. The major diffraction spots correspond to (121), (220), (202) and (240) of $LaFeO_3$ with perovskite structure. No diffraction spots were attributed to the secondary phase [16]. Thus, the morphological results analyzed using SEM and TEM images demonstrate that the flowers like structures are comprised of $LaFeO_3$ nanoplates.

(d) Microspheres Composed of Nanoparticles

In order to understand the morphology evolution of $LaFeO_3$ microspheres composed nanoparticles prepared by using lanthanum nitrate hexahydrate, ferric nitrate nonhydrate and citric acid as staring materials, several experiments were carried out with various hydrothermal reaction time of 6, 12, 18 and 24 h calcined at 800 °C and the corresponding SEM image are shown in Figure 3.6. The formation of microsphere was observed at 6 h, and few aggregated nanoparticles was also observed. Which proves the formation of $LaFeO_3$ microspheres were due to the self-assembly process of nanoparticles as shown in Figure 3.6a The SEM image for the increased reaction time of 12 h as shown in Figure 3.6b shows microspheres with smooth surface and diameters ranging from 2 to 3 μm with an average particle size of 70 nm as observed in TEM image shown in Figure 3.7b. There is no void and deficiency was found in these microspheres. This further implies that the synergistic effect of the cations and anions are necessary for the formation of $LaFeO_3$ microspheres. All these observations confirm the most plausible mechanism for the growth of the microspheres is that nanoparticles nucleate, grow, and aggregate. Subsequent increase in the reaction time to 18 h leads to the larger diameter of 3.5 - 4.5 μm as shown in Figure 3.6c and few aggregated small particles were found on the surface of the microspheres because of dissolution - recrystallization process which is often observed in the hydrothermal process [17]. Further the prolonged reaction time of 24 h leads to the destruction of single microspheres as shown in Figure 3.6d, resulting in combined microspheres which apparently involved the consumption of nanoparticles formed at 18 h of reaction time. It is indicating that the Ostwald ripening mechanism and reassembly process were operating in this

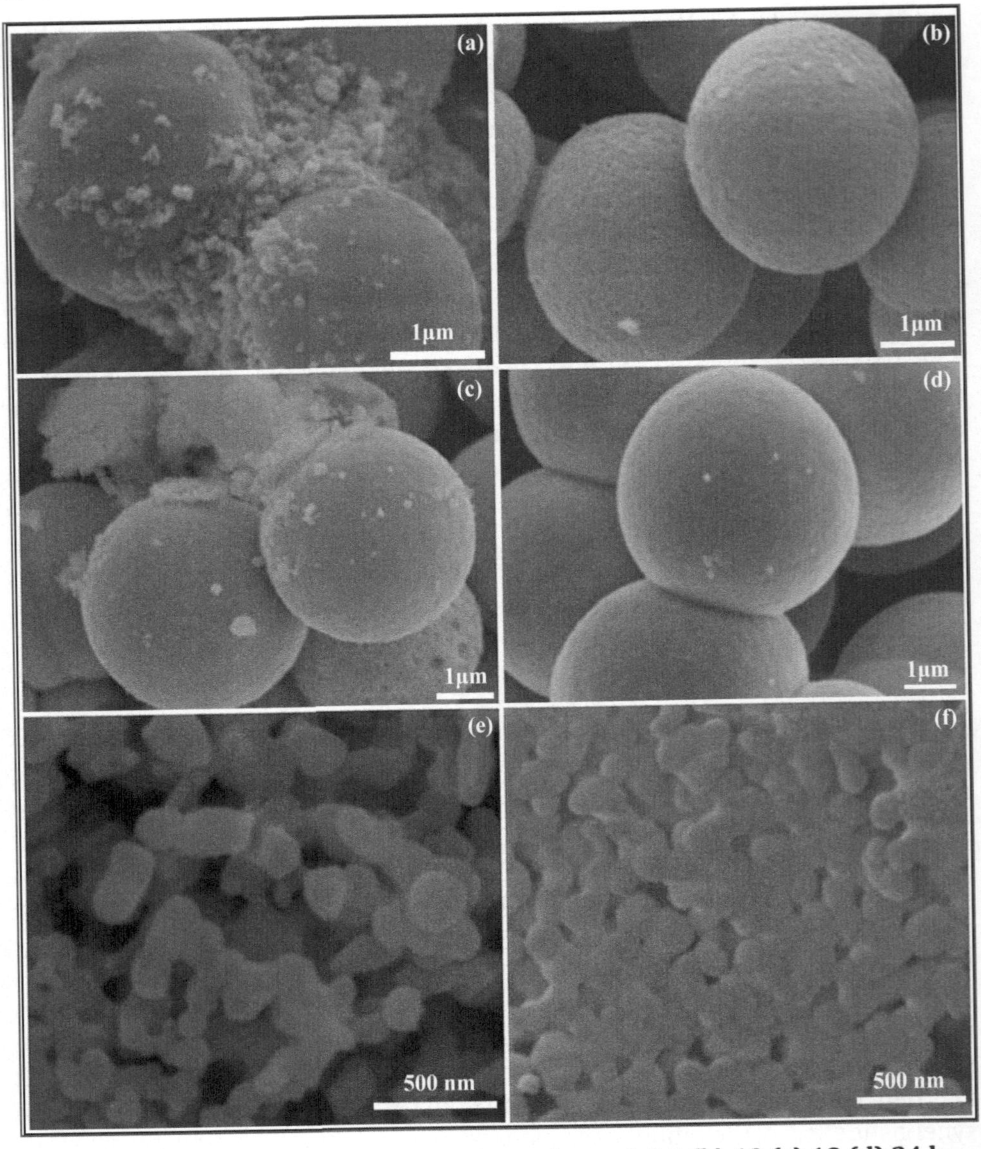

Figure 3.6. SEM Images of $LaFeO_3$ Microspheres (a) 6 (b) 12 (c) 18 (d) 24 h; High Magnification Image (e) 12 and (f) 24 h Hydrothermally Reacted at 180°C.

case. The surface of combined microspheres consists of particles with the average size of ~ 85 nm as clearly evidenced from TEM image shown in Figure 3.7d. On the basis of above discussions it is inferred that, in the initial reaction a group of freestanding crystallites with unequal sizes in nonequilibrium state were formed. To lower the total energy of the system, smaller crystallites would eventually dissolve into the solution and regrowth the larger ones during Ostwald ripening process [17]. These facts indicate that the formation of $LaFeO_3$ microsphere depends on the size of the primary particles and also on the reaction time. The ultra-fine particles formed at low reaction time benefits the formation of $LaFeO_3$ microspheres. On the whole, the formation of $LaFeO_3$ microspheres went through the process of nucleation, growth, self- assembly, dissolution and reassembly process. The diameter and the surface structure had no change when the reaction time was further increased and it suggest the process reached its steady-state under the experimental conditions employed.

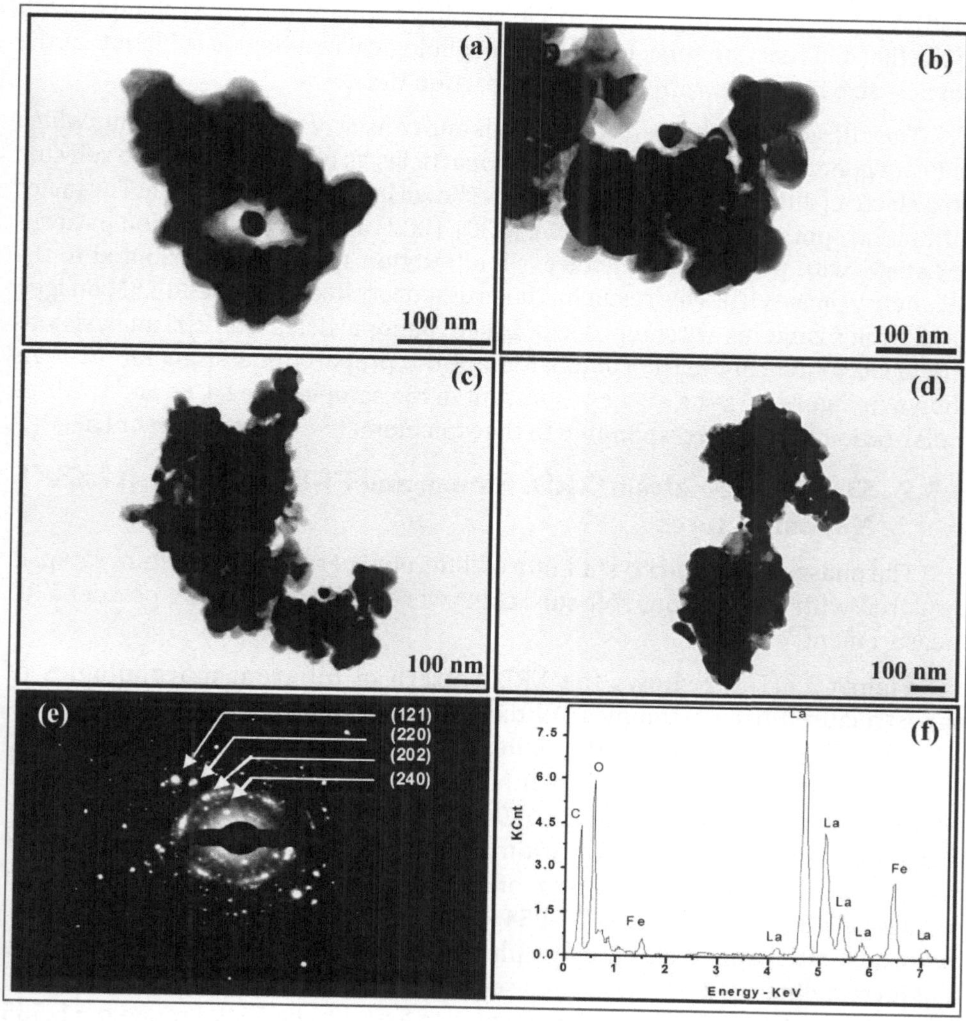

Figure 3.7. TEM Images of $LaFeO_3$ Nanoparticles (a) 6 h, (b) 12 h, (c) 18 h (d) 24 h and (e) SAED Pattern of 12 h and (f) EDAX Spectrum of 12 h.

TEM analysis of microspheres composed of $LaFeO_3$ nanoparticles prepared with the reaction time of 6 h reveals the nanoparticles of irregular shape with an average size ~ 60 nm as shown in Figure 3.7a, When the reaction time was increased to 12 h, sphere like nanoparticles with an average size ~ 70 nm as shown in Figure 3.7b. This elongation effect of nanoparticles was further evident when the reaction time was increased to 18 h as these nanoparticles undergo recrystallization process as shown in Figure 3.7c. It should be mentioned that nanoparticles on the microsphere are loosely bound for reaction time of 6 and 12 h when compared to 18 and 24 h. Figure 3.7d reveals the $LaFeO_3$ nanoparticles with an average size ~ 85 nm prepared at 24 h hydrothermal reaction time. TEM observation indicates that both the morphology and particle size have apparent regular changes with hydrothermal reaction time. From the morphological results it is evident that the particle size increases with increase in reaction time.

The diffraction rings are discontinuous and consist of rather sharp spots which indicate good crystallinity of $LaFeO_3$ nanoparticles as observed from the selected area electron diffraction (SAED) pattern as shown in Figure 3.7e for 12 h. The major diffraction spots correspond to (121), (220), (202) and (240) diffraction patterns of $LaFeO_3$ with perovskite structure. No diffraction spots were attributed to the secondary phase [18]. This result has been demonstrated by SEM and TEM images, *i.e.*, the microspheres are composed of $LaFeO_3$ nanoparticles. The EDS analysis was employed to determine the composition of the prepared materials for 12 h. As shown in Figure 3.7f, the elements present in the samples are La, Fe and O with a molar ratio of 1:1:3 corresponding to the stoichiometric composition of $LaFeO_3$.

3.3.2 Structural Analysis (XRD, Raman and FT-IR) of different $LaFeO_3$ Nanostructures

The phase, purity and crystallinity of different $LaFeO_3$ nanostructure samples prepared with and without soft surfactant were determined using powder XRD measurement.

Figure 3.8(i) a-f shows the XRD pattern of different morphologies of nanostructured $LaFeO_3$ samples. All the diffraction peaks for $LaFeO_3$ nanocubes shown in Figure3.8a was readily indexed to the perovskite phase with cubic structure. The interplanar spacing of (110) crystallographic plane matches well with the calculated lattice distance of 0.29 nm for nanocube from HRTEM image discussed in previous section. The estimated lattice constants were found to be $a = b = c = 3.929$ Å with a maximum error of ± 0.0003, which is in accordance with that of bulk $LaFeO_3$ crystals (JCPDS 75-0541). The XRD pattern of flowers, dendrites, nanorods, microspheres and nanospheres as shown in Figure 3.8b-f indicate that it crystallites as a perovskite phase with orthorhombic structure. The lattice parameters estimated from XRD pattern was found to be in the range $a = 5.565$ - 5.575, $b = 7.838$ - 7.849 and $c = 5.577$ - 5.587 Å, with a maximum experimental error of ± 0.0003 which are in good agreement with the literature value (JCPDS 37-1493) [19]. The density of $LaFeO_3$ nanostructures for all samples was estimated to be in the range 6.75 - 6.96 g cm^{-3} from XRD data [20]. The HRTEM image of $LaFeO_3$ nanostructures discussed in morphological analysis section shows clear lattice fringes along the growth direction. The spacing of the lattice fringes was 2.79 - 2.76

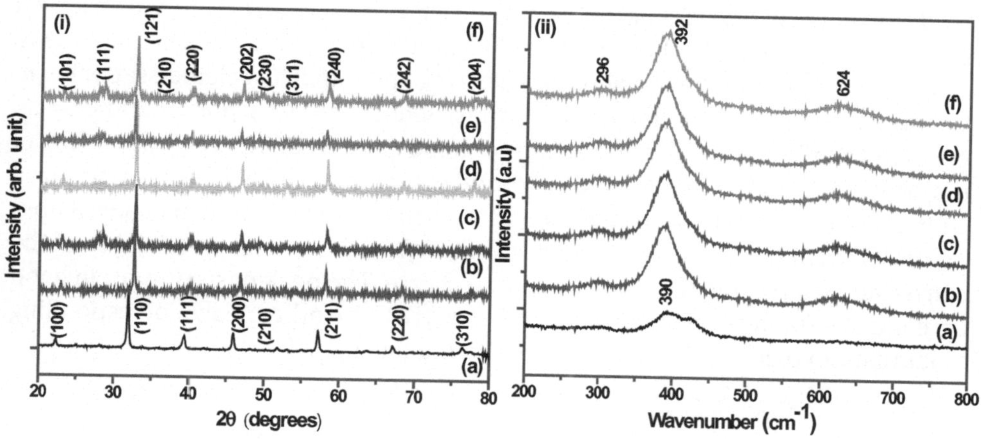

Figure 3.8. (i) X- ray Diffraction Pattern and (ii) Raman Spectrum of different $LaFeO_3$ Nanostructures (a) Nanocube, (b) Flowers, (c) Dendrites, (d) Nanorodsm, (e) Microspheres and (f) Nanosphere.

Å which was well indexed with the 'd' spacing of (121) plane of orthorhombic $LaFeO_3$ structure and the lattice parameters are almost the same for all nanostructures. No other diffraction peaks from La_2O_3/or Fe_2O_3 were observed. These results confirm the samples have transformed completely from cubic to orthorhombic structure using structure directing agents. The βcos θ value calculated from the Williamson - Hall plot was used in the Scherrer formula to calculate the average crystallite size and was found to be ~ 95, 88, 81, 75, 69 and 54 nm for the prepared nanocubes, flowers, dendrites, nanorods, microspheres and nanospheres respectively. These values are also in agreement with the results obtained from morphological analysis. In the hydrothermal growth conditions, nanoparticles of $LaFeO_3$ were formed and subsequently aggregated to give morphologies such as cube, rod and sphere. The cubic structure formation was obtained when structure directing agents or surfactants were not used. When structure directing agents like PVP, CTAB, urea and citric acid were used, the formed $LaFeO_3$ nanoparticles were of orthorhombic structure and these particles aggregate to form flowers, dendrites, rods and spheres-like morphologies. The surfactantions from PVP, CTAB, urea and citric acid play an important role in the formation of different morphologies.

Raman spectroscopy has the advantage of being very sensitive find the structure distortion and oxygen motion. So, a detailed investigation on electron excitation is very important for understanding in these perovskite-type materials [21]. Figure 3.8 (ii) a-f shows the room temperature Raman spectra recorded for all $LaFeO_3$ morphologies. Most of the vibrational modes are similar for orthorhombic structures, but few modes were not present for the cubic structure. We observed Raman peak for cube like $LaFeO_3$ structure at 390 cm^{-1} corresponding to B_{3g} assigned mode as shown in Figure 3.9a. In the case of orthorhombic structure, we observed similar mode as that of cubic structure. The modes at 296 and 624 cm^{-1} are due to A_g assigned modes and two - photon scattering and impurity related scattering in the case of orthorhombic flowers, dendrites, nanorods, microspheres and we also observed similar modes for orthorhombic nanospheres. Comparing the results of the stretching mode of cubic (390 cm^{-1}) and orthorhombic (392 cm^{-1}) structure, a shift in wavenumber towards higher side was observed as shown in Figure 3.9b-f. The shift could be due to local structural change in the crystal system, due to the conversion from cubic to orthorhombic structure. The peak positions are in good agreement with the values reported for $LaFeO_3$ [21]. No other peaks corresponding to lanthanum oxide and iron oxide were found like in XRD.

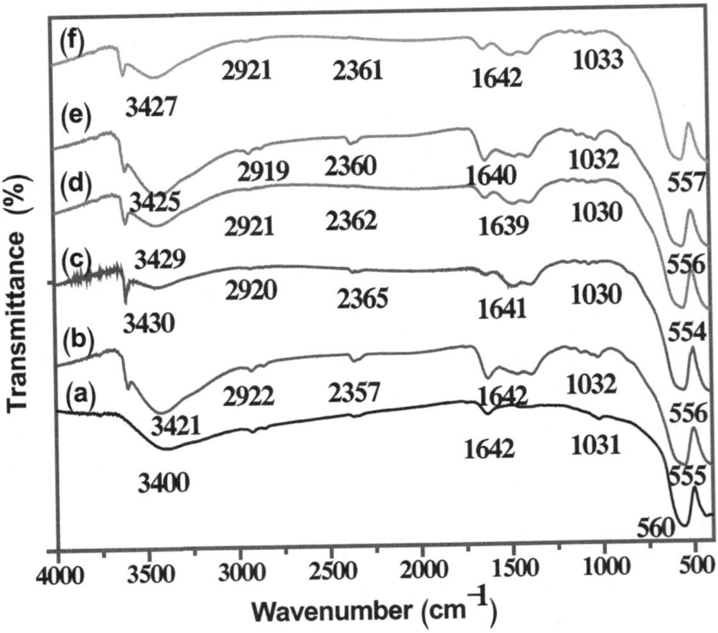

Figure 3.9. FT-IR Spectra of different Morphologies of $LaFeO_3$ Nanostructures (a) Nanocubes, (b) Flowers, (c) Dendrites, (d) Nanorods, (e) Microspheres and (f) Nanosphere.

Further, the FT-IR spectra were used to monitor and confirm the structure of different morphologies of $LaFeO_3$ nanostructures. The FTIR spectra shown in Figure 3.9 for different morphologies of $LaFeO_3$ nanostructures were taken to complete the information of long range order (XRD) with short range order (Raman and FTIR). The broad band observed at 3400 - 3429 cm^{-1} is characteristic of the H–O bending mode of absorbed water or the hydrogen bonded to the oxygen ions in the frame work. In addition, the band observed at 2922 - 2919 cm^{-1} correspond to the physically absorbed CO_2, and the two bands at 2357 - 2365 and 1639 - 1642 cm^{-1} originate from the symmetric and asymmetric stretching of carboxyl root.

The smaller peak at 1031 to 1033 cm^{-1} corresponds to the principal vibration of CO_3^{2-} group due to the exposure to ambient atmosphere, indicating that La-carbonate species exist on the $LaFeO_3$ nanostructures surface of perovskite compound which was not detected in the XRD. In addition, the intense band at 560 to 554 cm^{-1} can be attributed to Fe–O stretching vibration being characteristic of the octahedral FeO_6 group in $LaFeO_3$. Contrarily, no obvious characteristic peaks of surface adsorbed oxygen species were observed. The results infer that the prepared $LaFeO_3$ have carboxyl group, which is necessary for the development of functional application. The FTIR analysis thus gives a clear view of the various stages of transformation. It can be thus concluded that FTIR spectra of different $LaFeO_3$ nanostructures have similar structural units and formation of identical chemical moieties as reported in the literature [22]. This is also in agreement with the results obtained from XRD and Raman analysis.

3.3.3 Compositional Analysis of different $LaFeO_3$ Nanostructures

The elemental composition, purity and chemical status of different morphologies of $LaFeO_3$ nanostructures were carried out using X-ray photoelectron spectroscopy. The binding energies obtained from the XPS analysis were corrected with C 1s reference line at 284.6 eV. The typical XPS survey spectrum obtained from different $LaFeO_3$ nanostructures are shown in Figure 3.10a which reveals the presence of La(3d), Fe(2p), O(1s) and C(1s). The carbon peak can be attributed to adventitious carbon on the surface of the samples. The XPS La3d and Fe2p core-level spectra reveals that in $LaFeO_3$ samples, lanthanum and iron atoms are in the formal chemical valance state +3[23]. The high resolution spectra in Figure 3.10b shows the two strong La peaks were observed at 833 - 835 and 850 - 852 eV, which corresponds to spin - orbit splitting of $3d_{5/2}$ and $3d_{3/2}$ of La^{3+} ions in oxide form. The binding energy of Fe $2p_{3/2}$ and Fe $2p_{1/2}$ were observed at 708 - 709 and 723 - 724 eV as shown in Figure 3.10c which corresponds to Fe^{3+} ions in oxide form [24]. The results support that lanthanum oxide exists in the $LaFeO_3$ samples. The liberation of lanthanum oxide from $LaFeO_3$ perovskite lattice is due to the unusual oxidation state of iron, Fe^{4+} in the oxide. When a part of Fe^{3+} ions is oxidized to Fe^{4+} ions, a portion of lanthanum ions would be removed from the perovskite lattice based on the principle of controlled valency [24].The two O1s XPS spectra in Figure 3.10d are wide and asymmetric and it demonstrates that there are at least two kinds of O chemical states according to the binding energy at 528 - 528.5 eV ascribed to lattice oxygen species (OL) and 530 - 530.5 eV assigned to chemisorbed hydroxyl oxygen species (O^-) or OH^- species because the O(1s) binding energy of O^- or OH^- ion which is generally higher by 2.1 - 2.5 eV than that of lattice oxygen [25]. Thus, the O1s XPS spectrum is fitted to two kinds of chemical states by Gaussian rule. The binding energy at 528 - 528.5 eV, corresponding OL XPS signal can be attributed to the contribution of La–O and Fe–O in $LaFeO_3$ crystal lattice. The OH XPS signal at 530 - 530.5 eV can be related to the hydroxyl group resulting mainly from the chemisorbed water, since lanthanum oxide is easily hygroscopic when exposed to air. From the relative intensities of the XPS spectra, the compositional

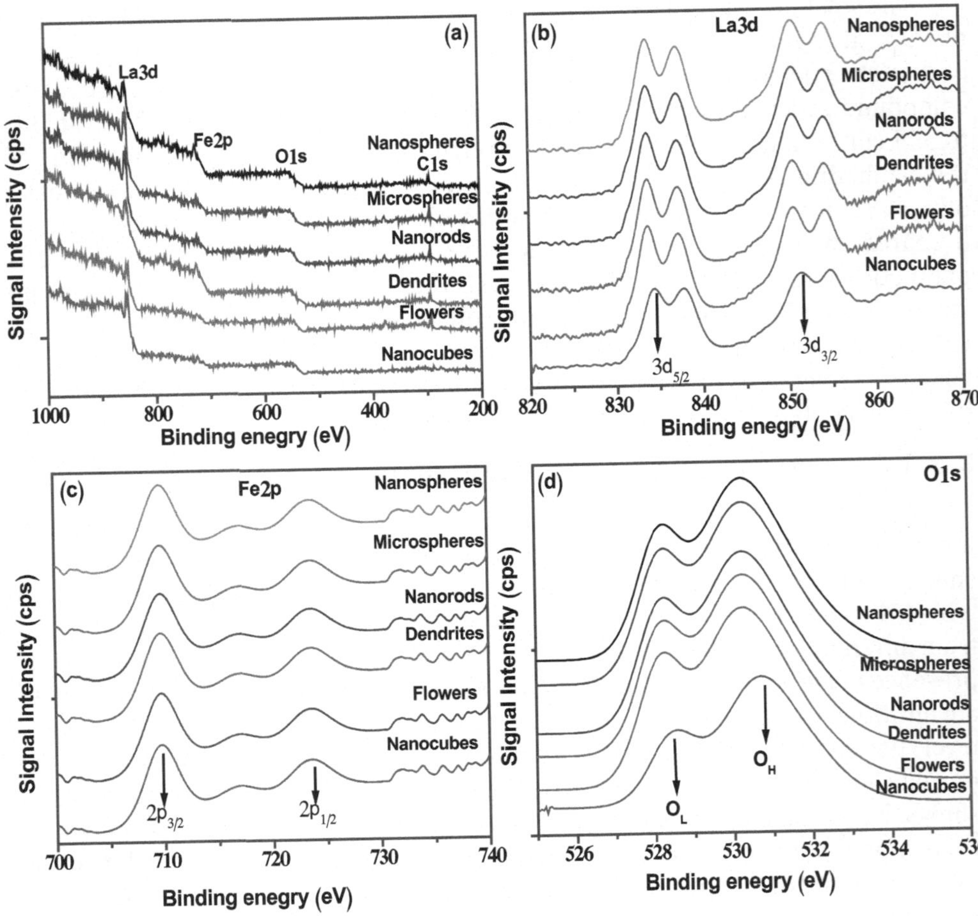

Figure 3.10. (a) XPS Survey Spectrum, (b) La3d, (c) Fe2p, and (d) O1s Spectrum of different $LaFeO_3$ Nanostrucutures.

stoichiometry was calculated between La, Fe and O and it was found to be 1:1:3. This kind of metal–support interaction has been taken as an important factor in the bio sensing, photocatalytic and gas sensor mechanism.

3.3.4 Thermal Analysis of different $LaFeO_3$ Nanostructures

Most of the perovskite oxide nanostructured materials may get dehydrated to some degree without major alternations in their crystal structure. Therefore, to evaluate the thermal stability of these synthesized $LaFeO_3$ nanostrucutres, the thermal analysis were performed. Thermogravimetric analysis of different morphologies of $LaFeO_3$ nanostrucutres was conducted from room temeapture (RT) to 1000 °C with a heating rate of 20 °C/min as shown in Figure 3.11a-f. In TG curve, initial weight loss of 2.5 - 4.0 wt per cent recorded in the temperature range RT - 295 °C and it can be attributed to the evaporation of physically absorbed

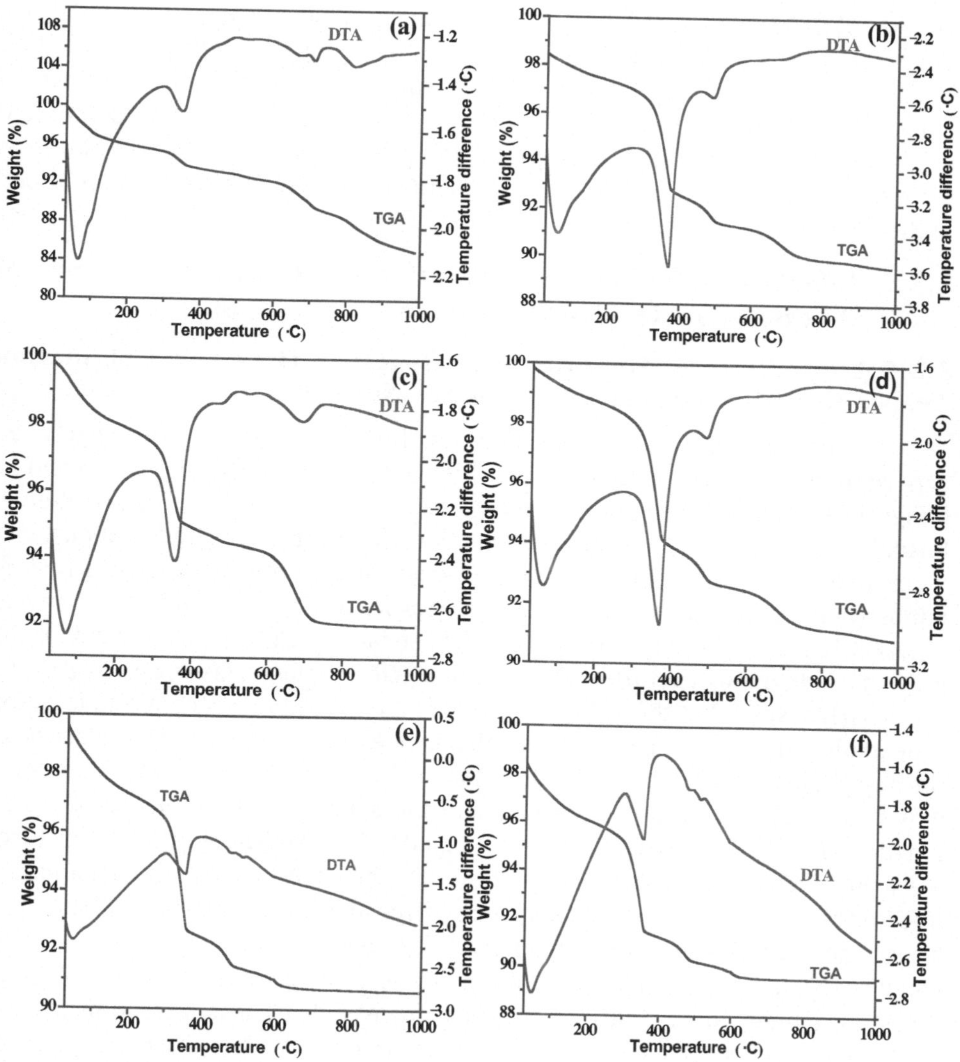

Figure 3.11. TG/DTA Analysis of different Morphologies of $LaFeO_3$ Nanostrucutures (a) Nanocubes, (b) Flowers, (c) Dendrites, (d) Nanorods, (e) Microspheres and (f) Nanosphere.

water which is also evidenced from the exothermic peak in the range 285 - 290 °C in differential thermal analysis (DTA) curve. As the temperature further increases, a notable weight loss of 4.2 to 5.5 per cent is observed between 295 and 375 °C, at the same time the DTA curve shows an endothermic peak in the temperature range 365 to 375 °C, which can be assigned to the decomposition of nitrates and other organic impurities [26]. A minor weight loss of 2.1 - 2.5 per cent observed in the temperature in the range of 380-700 °C can be assigned to the complete decomposition of oxycarbonates. A small exothermic peak observed in the range of 590 - 610 °C can be possibly attributed to the gradual crystallization of $LaFeO_3$. There was no weight loss detected by further increasing the temperature to 1000 °C. The TG/DTA curves gave strong evidence that the proposed molecular formula $LaFeO_3$ was rational. These observations are comparable to the results obtained from the XRD pattern of the as prepared and calcinated samples which confirms the transformation of amorphous to crystalline phase which have been discussed in the following sections.

3.3.5 Effect of Surfactant Concentration and Growth Parameters of different $LaFeO_3$ Nanostructures

3.3.5.1 (a) Effect of Hydrothermal Reaction Time and Calcination Temperature of Nanocubes, Nanorods and Nanospheres

In order to study the morphology evolution of $LaFeO_3$ nanostructures, experiments were carried out at various hydrothermal reaction time of 3, 6 and 12 h Figure. 3.12 shows the nanostructure (nanocubes, nanorods and nanospheres) formation occurs at 3 h hydrothermal reaction. However aggregated nanoparticles were also observed in the micrograph which proves the formation of $LaFeO_3$ nanostructures are due to the self assembly process of nanoparticles as shown in Figure 3.12a, c, e. These observations, infer the most plausible mechanism for the growth of the nanostructure is that nanoparticles nucleate, grow, and aggregate. It is clearly visible that the growth occurs for the subsequent increase in reaction time to 6h and also few aggregated small particles were found on the surface of all the morphologies in Figure 3.12 b, d, f), due to dissolution-recrystallization process which is often observed in the hydrothermal synthesis of nanomaterials [27]. Further when the reaction time was prolonged to 12 h well defined nanostructures were observed as shown in Figure 3.12 a-c. On the basis of above discussions it is inferred that, in the initial reaction a group of freestanding crystallites with unequal sizes in nonequilibrium state were formed. To lower the total energy of the system, smaller crystallites would eventually dissolve into the solution and regrowth takes place ones during Ostwald ripening process [13].These facts indicate that the formation of different $LaFeO_3$ morphology depends on the size of the primary particles and also on the reaction time. On the whole, the formation of $LaFeO_3$ nanostructures such as nanocubes, nanorods and nanospheres went through the process of nucleation, growth, self-assembly, dissolution and recrystallization. The morphology and the surface structure had no change when the reaction time was further increased and it suggest the process reached its steady-state under the experimental condition employed.

3.3.5.1 (b) Growth Mechanism of $LaFeO_3$ Nanostructures

The growth mechanism of different $LaFeO_3$ nanostructures can be determined by both the internal and external conditions such as molar concentration, temperature and time. It is well known that the shape of nanoparticles can be controlled by structure directing agents. The formation process of $LaFeO_3$ nanocubes, nanorods and nanosphers is speculated in Figure 3.13.

Figure 3.12 SEM Image of Three different $LaFeO_3$ Morphologies (a) 3h, (b) 6h for Nanocubes, (c) 3h, (d) 6h for Nanorods, (e) 3h and (f) 6h for Nanospheres.

During the initial period of hydrothermal reaction, metal ions in the solution react without and in the presence of urea and citric acid to form relatively stable distorted primary particles of $LaFeO_3$ under supersaturated condition. This decreases the free metal ion concentration in solution and results in slow generation of primary particles. The formed $LaFeO_3$ nanoparticles aggregate together which is driven by the minimization of interfacial energy [28]. The aggregated nanoparticles would act as primary cores that favour the homogenous assembly to produce larger self-assemble aggregates via self-assembly mechanism. The nanoparticles self-assemble along three-dimensional direction to form nanocubes. With prolonged hydrothermal reaction time dissolution-recrystallization process takes place where the smaller $LaFeO_3$ nanoparticles disappear and further it grows into $LaFeO_3$ cubes.

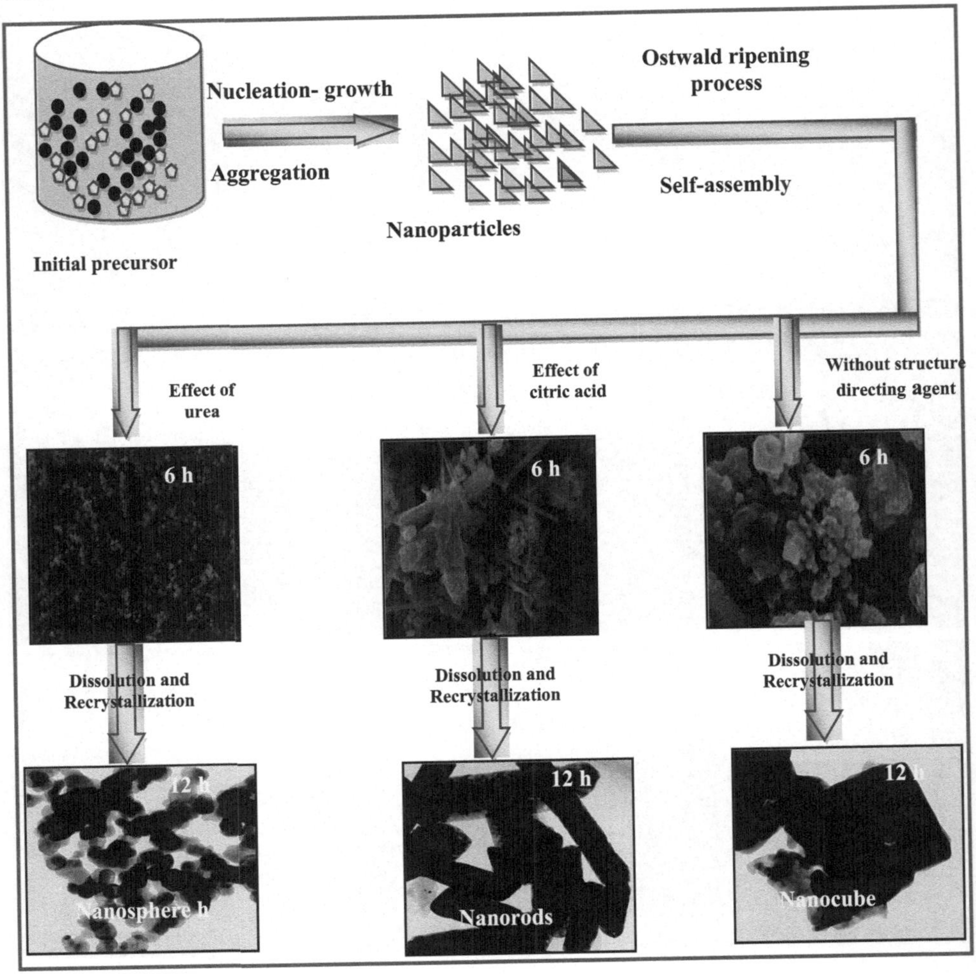

Figure 3.13. Schematic Illustration of the Growth Process of different $LaFeO_3$ Nanostructures.

The transition from cube to rod like morphology of $LaFeO_3$ particles occur with the addition of OH^- ions (via-urea) into the solution. Decomposition of urea (into NH_4OH and H_2CO_3) occurs during the hydrothermal reaction process followed by dissociation of the reaction product to yield OH^- and CO_3^{2-} ions. The gradual release of OH^- ions results in the nucleation and growth of uniformly sized rod like nanostructures. The hydrogen atoms of urea form strong hydrogen bridges with the oxygen atoms of $LaFeO_3$, thereby stabilizing the structure as a whole. Another transition of the nanocubes to sphere like morphology takes place when citric acid mediated OH^- ions get absorbed on the surface of $LaFeO_3$ nanocubes which directs the self-assembly of nanoparticles by molecular interaction into nanospheres with certain crystallographic orientation. Because sphericity has minimal surface energy driven by the minimization of total energy of the system, nanoparticles would be dissolved and recrystallized into nanospheres under hydrothermal growth process. The formation mechanisms of the three different $LaFeO_3$ nanostructures are similar. Owing to the effect of the surfactants, different morphologies are self-organized. The surfactants as structure directing agents have played a key role in the formation of these different nanostructures.

3.3.5.2 (a) Impact of Calcination Temperature and CTAB Concentration in the Formation of $LaFeO_3$ Dendrites Nanostructures

The powder XRD pattern for the as-prepared $LaFeO_3$ samples and those calcinated at different temperatures (600, 700 and 800 °C) are shown in Figure 3.14. The as-prepared sample shows amorphous nature and is characterized by the broad continuum peaks. The transformation of amorphous phase to crystalline $LaFeO_3$ was not completed at 600 °C as it shows the presence of only few characteristic peaks of $LaFeO_3$ with low intensity. When the calcination temperature was increased, the onset of $LaFeO_3$ crystallization can be observed but with low intensity peaks at 700 °C. Increasing the calcination temperature to 800 °C, high crystalline quality of perovskite type $LaFeO_3$ samples were revealed. All the diffraction peaks can be indexed to the orthorhombic perovskite structure. This is also in agreement with the results obtained from thermal analysis.

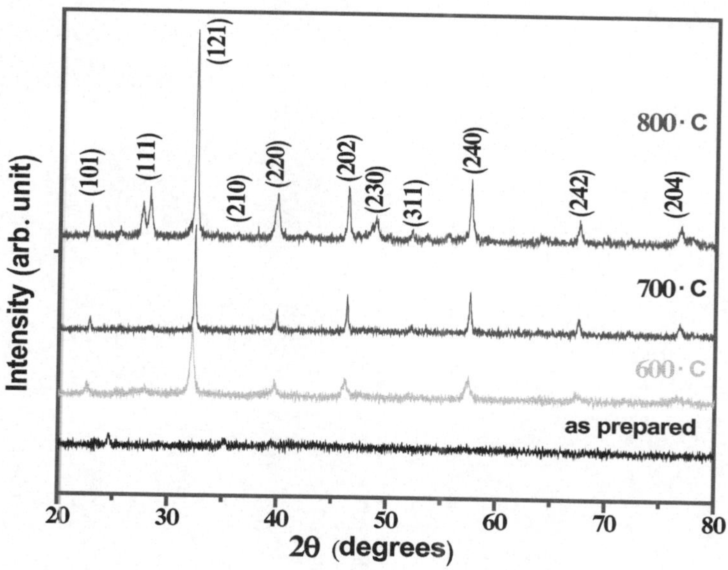

Figure 3.14 XRD Pattern of Nanostructured $LaFeO_3$ for different Calcination Temperatures.

The surface morphology of the prepared $LaFeO_3$ samples was found to be different depending on the calcination temperature and CTAB concentration as shown in Figure 3.15. Agglomerated nanoparticles were observed for the as-prepared

$LaFeO_3$ samples as in Figure 3.15a. The formation of dendritic nanostructures with primary order branches started to grow at 600 °C calcination, as observed in Figure 3.15b. When the calcination temperatures increases, dendrites via self-assembly process were observed at 700 °C, but crystallinity was not achieved as shown in Figure 3.15c. Further increase the calcination temperature to 800 °C shows clear morphology with well-defined dendritic structures containing main stem and branches were observed as in Figure 3.15d. The above results indicate that the calcination temperature is an important parameter in the formation of morphology and structure. To analyze the effect of CTAB surfactant on the final formation of nanostructure $LaFeO_3$ dendrites, we also have examined the influence of CTAB

Figure 3.15. SEM Image of (a) as Prepared, (b) 600 °C, (c) 700 °C, (d) 800 °C, (e) Low Concentration, and (f) High Concentration CTAB of $LaFeO_3$ Dendritic Nanostructures.

concentration on morphology of $LaFeO_3$ in the hydrothermal process. The molar concentration of La: Fe: CTAB used were (1:1:1), (1:1:2), and (1:1:3) respectively. Aggregated nanoparticles were observed when (1:1:1) molar concentration was employed in the typical synthesis process as shown in Figure 3.15e.

While with higher concentration of CTAB (1:1:3), more particles were arrested by the surfactant molecules leaving behind very few particles for the dendritic formation. Thus, the use of excess surfactant hindered the formation of dendritic structures resulting in spherical and incompletely formed branches as shown in Figure 3.15f. Clear morphology and smooth surface of $LaFeO_3$ dendrites were observed in Figure 3.3b when the La: Fe: CTAB molar concentration of (1:1:2) in the reaction solution, which indicates that appropriate concentration of CTAB is a crucial prerequisite for the formation of perfect dendrites under selected hydrothermal reaction conditions. The absorption of CTAB would direct the self-assembly of nanoparticles by molecular interaction into dendrites with certain crystallographic orientation. However, in the optimum precursor concentration, the diffusion is limited and under this diffusion limited regime the growth proceeds further via oriented attachment of particles. The oriented attachment of particles starts with a seed particle formed by nucleation under supersaturation and the growth of dendrites proceeds from this seed particle by the sticking and stacking of the diffusing particles at certain favourable sites determined by the crystal structure. This result is consistent with the observation that surfactant assisted reaction is a simple and effective way to control the nucleation and growth process [29]. Finally, all the above results confirm that the calcination temperature finds importance for controlled growth of dendrite structure. Hence, single phase $LaFeO_3$ dendrite perovskite structures with high crystalline quality with clear morphology were formed at 800 °C with CTAB concentration (1:1:2) was used for further analysis and biosensing application in the present work.

3.3.5.2 (a) Growth Mechanism and Influencing Factors of $LaFeO_3$ Dendrite Nanostructures

The growth mechanism of $LaFeO_3$ dendritic nanostructures can be determined by both the internal and external conditions such as calcination temperature and surfactant concentration. It is well known that the shape of nanoparticles can be controlled by surfactant and the possible formation mechanism is as follows:

$$La(NO_3)_3 \longrightarrow La^{3+} + 3NO_3^- \quad (1)$$

$$[Fe(CN_6)]^3 \longrightarrow Fe^{3+} + 3OH^- \longrightarrow Fe(OH)_3 \quad (2)$$

$$La^{3+} + C_{19}H_{42}BrN \longrightarrow [C_6BrNH_{13}]^{3-} La^{3+} + 3H^+ \quad (3)$$

$$[C_6BrNH_{13}]^{3-} La^{3+} + 3OH^- \longrightarrow La(OH)_3 + [C_6BrNH_{13}]^{3-} \quad (4)$$

$$La(OH)_3 + Fe(OH)_3 \longrightarrow LaFeO_3 + 3H_2O \quad (5)$$

La^{3+} ions were produced when $La(NO_3)_3$ was dissolved in water (Eq. (1). $[Fe(CN_6)]^3$ dissociate into Fe^{3+} ions and further hydrolyze to form $Fe(OH)_3$ under hydrothermal reaction temperature as they are highly stable at room temperature (Eq. (2) [30]. La^{3+} combines easily with CTAB via the chelating effect than Fe^{3+} owing to the much higher stability constant (Eq. (3). La^{3+} will be released slowly and gradually converted into $La(OH)_3$ via the alcoholysis reaction during hydrothermal process (Eq. (4). The simultaneously formed $Fe(OH)_3$ react with $La(OH)_3$ to form the $LaFeO_3$ precipitates by the condensation effect (Eq. (5). The shape of the precursor plays a major role on the final formation of dendritic $LaFeO_3$ structures. It is known that CTAB not only accelerates the reaction speed of the growth units, but also directs their oriented attachment, growth of diffusion- limited aggregation.

The formation process of $LaFeO_3$ dendritic nanostructures is speculated in Figure 3.16. The reaction temperature and precursor concentration significantly influences the morphology of the product which has been discussed in the previous section 3.3.5.2. Nanostructured dendrites were obtained under hydrothermal optimized conditions as in Figure 3.15. These results suggest that controlled size and shape of $LaFeO_3$ dendritic nanostructures can be obtained by controlling the kinetic parameters of the reaction process. At the initial stage, the slow dissociation nature of ions in the precursor solution and the controlled diffusion of $LaFeO_3$ arising due to temperature and concentration gradient under hydrothermal conditions also play an important role in the growth of dendritic structures. The nanoparticles formed by nucleation under supersaturation condition serve as an immobile seed in the solution. The free nanoparticles from a random position far away diffuse towards the seed.

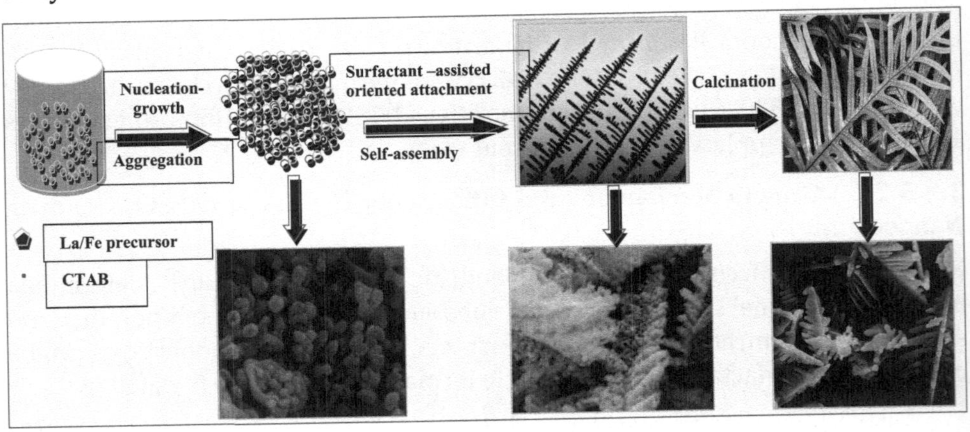

Figure 3.16. Schematic Diagram Illustrating the Growth Mechanism of $LaFeO_3$ Dendritic Nanostructures.

Once they touch the seed, they are immobilized instantly on the empty surface of the seed yielding an initial aggregate of $LaFeO_3$ nanoparticles as shown in Figure 3.3c, with this seed as the centre and the attached nanoparticle as the arm. With elapsed time, other free nanoparticles will diffuse continuously towards the aggregate and further would be immobilized on the empty surface of the arm forming larger aggregates. This crystal growth process continues until all free nanoparticles are depleted, leading to dendritic nanostructure. Actually, a large mass of seeds are distributed uniformly in the solution and multiple aggregation events occur simultaneously in the respective regions of the solution. As a result, dendrites with limited size were obtained. The physical adsorption of diffusing $LaFeO_3$ crystals when contacted with each other leads to the formation of symmetrical dendrites. The growth of dendrites is globally diffusion-controlled but locally accomplished by oriented attachment, which was found to be responsible for the formation of dislocations during early crystal growth [31].

In equilibrium condition, the crystals grown under hydrothermal conditions may exhibit simple morphologies with minimum surface energy. In our case, as the system is driven farther from equilibrium, the growth rate of crystals exceeds the mass transport rate of ions that feed the growing crystal leading to the formation of main stem and symmetric branches on both sides called dendrites were promoted. Therefore, it is reasonable to suggest that rapid nucleation and growth kinetics of $LaFeO_3$ nanoparticles contribute mainly to the formation of dendrites. The controlled growth of dendrites system by diffusive process can be related to the diffusion - limited aggregation (DLA) model developed to interpret this type of system [32]. The rate of nucleation and subsequent growths of dendritic nanostructures are effectively controlled by the hydrothermal reaction condition and surfactant concentration [32]. This result is consistent with the observation that surfactant assisted reaction is a simple and effective way to control the nucleation and growth process. The surfactant CTAB as a structure directing agent has played a key role in the formation of $LaFeO_3$ dendritic nanostructures.

3.3.5.3 (a) Effect of Hydrothermal Condition and Surfactant Concentration of Flower like $LaFeO_3$ Nanostructures

The influence of hydrothermal condition and surfactant concentration on the flower-like $LaFeO_3$ nanostructures composed of nanoplates were systematically studied from the SEM images. In the present analysis, we find that the reaction temperature is undoubtedly play a vital role in the nucleation and growth of $LaFeO_3$ crystallites which determines the final flower like morphology of the products. The obtained samples have been investigated in detail to reveal the relationship between morphology, reaction time and calcination temperature and surfactant concentration. Agglomerated nanoparticles were observed for the as-prepared samples as shown in Figure 3.17a. When the powders were calcined at 600 °C agglomerated nanoparticles were involved in the early stages of flower like morphology formation as in Figure 3.17b. Further increasing the calcination temperature to 800 °C as in Figure 3.5a, clear morphology of flower like $LaFeO_3$ nanostructure were observed. Therefore, the results indicate that the calcination temperature was an important parameter that controlled the morphology and structure. As shown in Figure 3.17c, at an early reaction stage for 3h, many nanoparticle aggregates were formed and careful observation demonstrates that some small nanoplate bundles grow out of the surface of spherical core. When the

Figure 3.17. SEM Images of $LaFeO_3$ Nanostructures (a) as Prepared, (b) 600 °C, (c) 3h, (d) 6h, (e) Low PVP Concentrations, (f) High PVP Concentration.

reaction time was prolonged to 6 h, the size of the spherical core became smaller than the samples reacted for 3h. Meanwhile, the nanoplates on the surface of the spherical core obviously became longer and their diameter also increased. Bundles of underdeveloped flower like architectures constructed by many nanoplates emerged as shown in Figure 3.17d. Further increasing the reaction time to 12h, clearly reveals the spherical core completely disappeared and typical 3D flowerlike microstructures comprised of nanoplates were formed as shown in

Figure 3.5a. Moreover, increasing the reaction time to 15 h, the morphology of the products did not change remarkably. In the present system, it was found that the amount of PVP also plays a key role in the formation of nanoplates built flower like architectures. Figure 4.18e-f shows SEM images of the product prepared with different molar amounts of La: Fe: PVP *viz.* (1:1:1), (1:1:2) and (1:1:3) respectively. Aggregated nanoparticles were observed when (1:1:1) molar concentration was employed in the typical synthesis process as shown in Figure 3.18e. While with higher concentration of PVP (1:1:3), the yield of nanoparticles became higher and irregularly aggregated flower like morphology were obtained (Figure 3.18f). Well arranged and uniform flower like morphology as shown in Figure 3.5b were formed when the La: Fe: PVP molar concentration was 1:1:2 in the reaction solution, which indicates that appropriate concentration of PVP is a crucial prerequisite for the formation of perfect flower like morphology under selected reaction conditions. This result is consistent with the observation that surfactant assisted reaction is a simple and effective way to control the nucleation and growth process [33]. The morphology evolution process mentioned above indicates that a suitable temperature and surfactant concentration can effectively control the production rate of $LaFeO_3$ nanoparticles and thus kinetically control the aggregation and growth of $LaFeO_3$ crystallites.

3.3.5.3(b) Possible Formation Process of Flower like $LaFeO_3$ Nanostructure

To understand the growth mechanism of flower like $LaFeO_3$ nanostructure we have investigated the morphology evolution with respect to different reaction time, as shown in Figure 3.18. Based on the obtained results, it is well known that the shape of nanostructures can be controlled by surfactant and the possible reaction mechanism is as follows

$$La(NO_3)_3 \longrightarrow La^{3+} + 3NO_3^- \quad (1)$$

$$[Fe(CN_6)]^3 \longrightarrow Fe^{3+} + 3OH^- \longrightarrow Fe(OH)_3 \quad (2)$$

$$La^{3+} + (C_6H_9NO)x \longrightarrow [C_6H_6NO]^{3-} La^{3+} + 3H^+ \quad (3)$$

$$[C_6H_6NO]^{3-} La^{3+} + 3OH^- \longrightarrow La(OH)_3 + [C_6H_6NO]^{3-} \quad (4)$$

$$La(OH)_3 + Fe(OH)_3 \longrightarrow LaFeO_3 + 3H_2O \quad (5)$$

La^{3+} ions were produced when $La(NO_3)_3$ was dissolved in water (Eq. (1). $[Fe(CN_6)]^3$ dissociate into Fe^{3+} ions and further hydrolyze to form $Fe(OH)_3$ under hydrothermal reaction temperature as they are highly stable at room temperature (Eq. (2) [30]. La^{3+} ions combine easily with PVP via the chelating effect than Fe^{3+} owing to the much higher stability constant than Fe^{3+} complex (Eq. (3). La^{3+} ions will be released slowly and gradually converted into $La(OH)_3$ via the alcoholysis reaction during hydrothermal process (Eq.(4). The simultaneously formed $Fe(OH)_3$ react with $La(OH)_3$ to form the $LaFeO_3$ precipitates by the condensation effect (Eq. (5). The formation process of flower like $LaFeO_3$ nanostructures is speculated in Figure 3.18. The reaction time and precursor concentration significantly influences the morphology of the product which has been discussed earlier. From the obtained results it is suggested that the controlled size and shape of flower like $LaFeO_3$ nanostructures can be obtained by controlling the kinetic parameters of the reaction process. At the very beginning, through the interfacial reaction of $La(NO)_3$, $K_3[Fe(CN)_6]$ and PVP precursors, aggregated nanoparticles were produced which quickly grew into small sized nanoplates under the hydrothermal conditions.

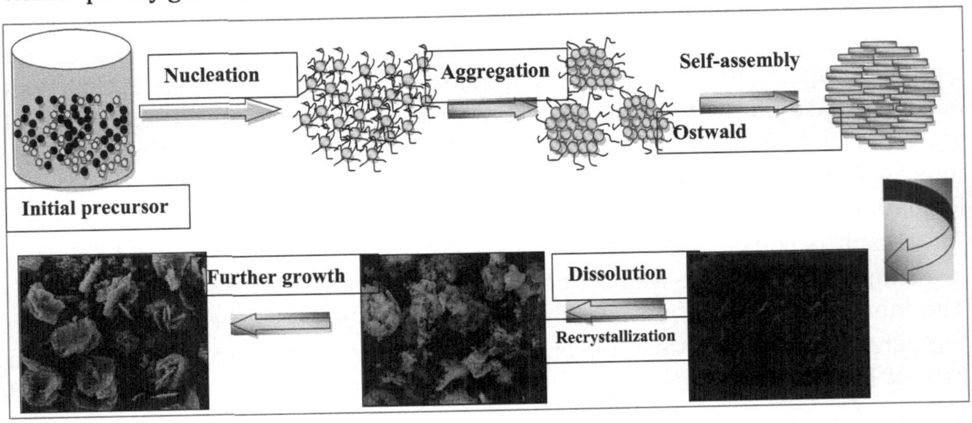

Figure 3.18. Schematic Representation of Formation Mechanism of flower like $LaFeO_3$ Nanostructures.

As a capping agent, polymer PVP can be easily adsorbed on the surface of the nanoparticles due to its large number of carbonyl groups and amide ligands. These nanoparticles are highly mobile at the interface and quickly aggregate into bigger spherical aggregates to decrease the surface energy. Assembly of the particles at liquid-liquid interface was also driven by a decrease in the interfacial energy. After the primary nanoparticles spontaneously aggregated together to form bigger aggregations at a faster rate under initial high supersaturation, the particles on or near the outermost surface of the aggregates would continue to grow as nanoplates at a slow rate under low supersaturation condition. Due to the high surface energy of the smaller nanoplates compared to the larger ones, the smaller nanoplates dissolve easily and diffuse outwards. The large nanoplates located on the surfaces would serve as growth sites for the subsequent recrystallization process and continued crystal growth. Because of the steric effect of PVP molecules, the aggregated nanoparticles are loosely compacted together with a spherical appearance and many voids exists in the interior of the aggregates. During the dissolution and recrystallization process, the voids in the interior of the spherical aggregates will provide the channels for mass transportation. At the earlier stage of crystal growth, PVP as potential crystal face inhibitors may be selectively adsorbed onto specific planes, which promotes the anisotropic growth of nanoplates [34]. Due to the gradual mass diffusion from the inner spherical core to the outermost surface, the nanoplates located on the outer surface grow in length and width at the expense of the dissolved core materials. Finally, after prolonging the reaction to 12 h, the spherical cores are consumed completely and the typical flowers like architectures constructed by several nanoplates were formed. In addition to the effect of the amount of PVP on the growth of nanostructures, the reaction time and PVP concentration would change the interfacial reaction rate by controlling the diffusion of reactant ions across the interface, which can regulate the kinetics of nucleation and growth of products and further efficiently control the morphology and structure of the final products.

3.3.5.4 (a) Effect of Calcination Temperature and Citric Acid Concentration on the Formation of $LaFeO_3$ Microspheres Composed of Nanoparticles

The XRD pattern of as-prepared $LaFeO_3$ samples and those calcinated at different temperature are shown in Figure 3.19. The as-prepared sample shows amorphous nature and it is characterized by the broad continuum peaks. The transformation of amorphous phase to crystalline $LaFeO_3$ was not completed at 600 and 650°C as the XRD pattern shows the presence of only few characteristic peaks of $LaFeO_3$ with low intensity. When the calcination temperature was increased the onset of $LaFeO_3$ crystallization can be observed but with low intensity peaks at 700 and 750 °C. Increasing the calcination temperature to 800 °C, well crystallized perovskite type $LaFeO_3$ (JCPDS 37-1493) with orthorhombic structure was obtained. The sharp and narrow diffraction peaks at 800 °C reveal that the samples have relatively high crystalline quality than those calcined at lower temperatures. This

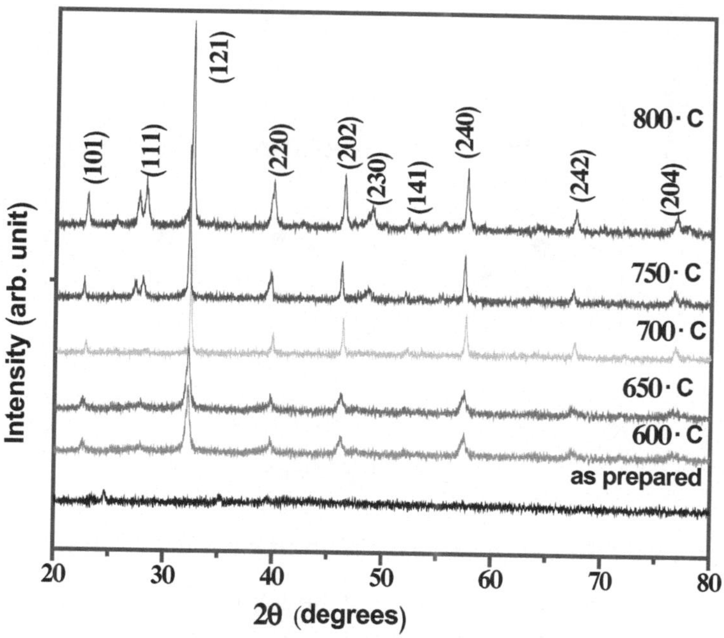

Figure 3.19. (a) XRD Pattern of $LaFeO_3$ Nanoparticles for different Calcination Temperature.

is also in agreement with the results obtained from thermal analysis. The intensity of diffraction peaks increased with increasing calcination temperature indicates the growth and crystallinity of $LaFeO_3$ microspheres as observed in Figure 3.20a. Figure 3.20b shows the XRD pattern for the different concentration of citric acid study the effect of concentration on the purity of prepared $LaFeO_3$ samples can be studied. Pure perovskite $LaFeO_3$ oxides were obtained when the metal ion to citric acid concentration was maintained at 1:1:2. On further increasing the concentration to 1:1:3 presence of impurity like La_2O_3 was noticed. When the amount of citric acid was reduced and the molar ratio maintained at 1:1:1 impurity like La_2O_3, Fe_2O_3 and $La(OH)_3$ were present, but $LaFeO_3$ products could not be observed indicating the incomplete decomposition of the product.

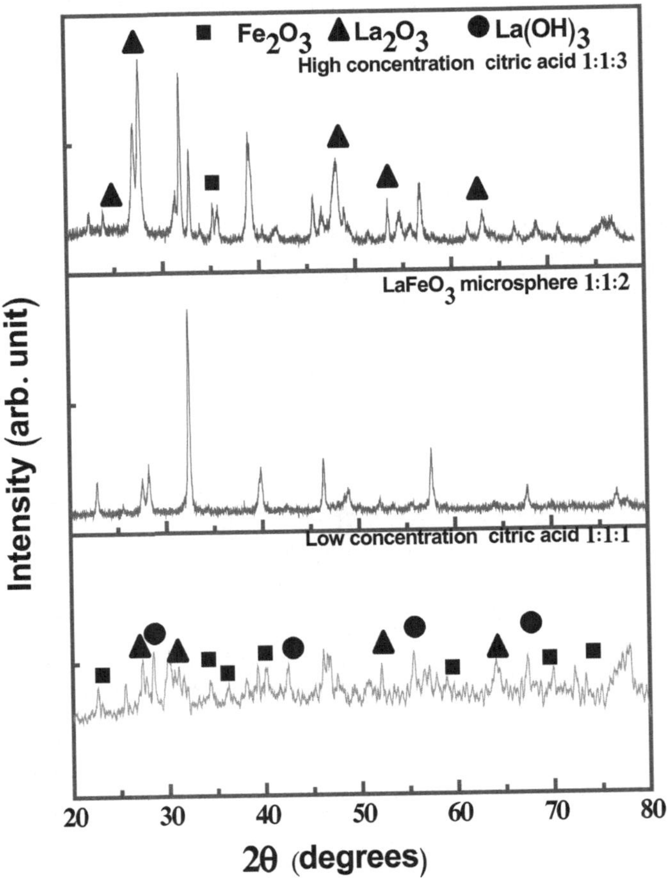

Figure 3.20 (b) XRD pattern of different citric acid concentration

The scanning electron micrographs of the as-prepared $LaFeO_3$ powder and for different calcination temperatures were also analyzed. Agglomerated nanoparticles were observed for the as-prepared samples as in Figure 3.21a. When the powders were calcinated at 600 °C agglomerated nanoparticles were completely converted to microspheres via self-assembly process as shown in Figure 3.21b. Further increasing the calcination temperature to 800 °C in Figure 3.6b, clear morphology of $LaFeO_3$ microspheres were observed. Therefore these results indicate the calcination temperature was another important parameter that controlled the morphology and structure.

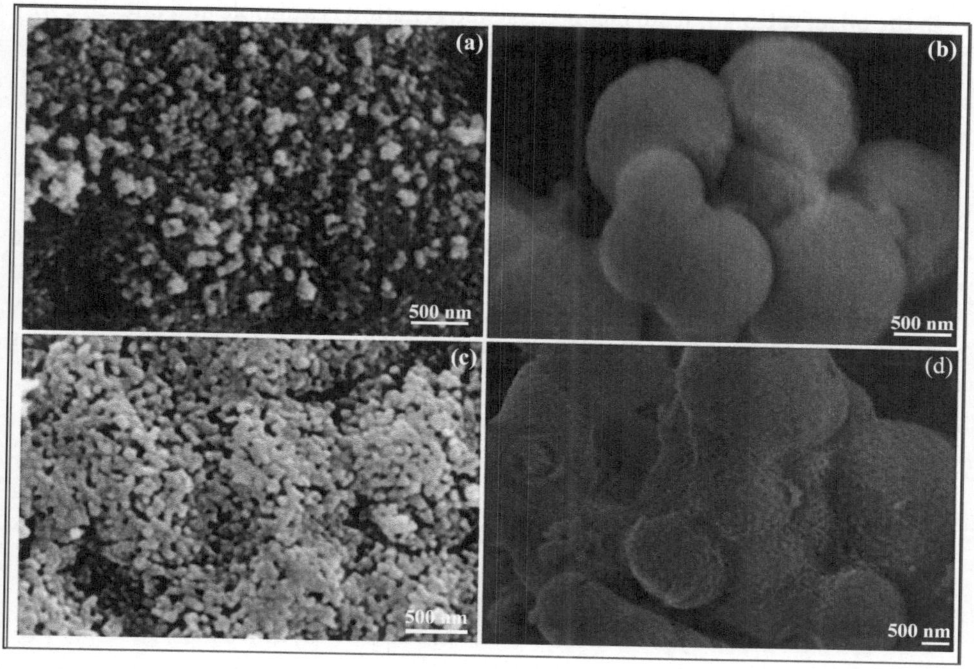

Figure 3.21. SEM Images of $LaFeO_3$ Microspheres (a) as Prepared, (b) 600 °C, (c) Low Citric Acid Concentaration, (d) High Citric Acid Concentration.

To analyze the effect of citric acid (CA) on the final form of nanostructure $LaFeO_3$ microspheres, we also examined how the citric acid concentration influenced the morphology of $LaFeO_3$ in a hydrothermal process with different concentrations of CA. The molar concentration of La: Fe: CA used were (1:1:1), (1:1:2), and (1:1:3) respectively. Aggregated nanoparticles were observed when (1:1:1) molar concentration was employed in the typical synthesis process as shown in Figure 3.21c. While with higher concentration of CA (1:1:3) the yield of nanoparticles became higher and irregularly aggregated microspheres of diameter about 2-3 μm were obtained as in Figure 4.21d.

Smooth surfaced single microspheres as shown in Figure 3.6b, were formed when the La: Fe: CA molar concentration was 1:1:2 in the reaction solution, which indicates that appropriate concentration of CA is a crucial prerequisite for the formation of perfect microspheres under selected reaction conditions. The citric acid absorbed on the surface of $LaFeO_3$ microsphere would direct the self-assembly of nanoparticles by molecular interaction into microsphere with certain crystallographic orientation. Because sphericity has minimal surface energy driven by the minimization of the total energy of the system, the microsphere composed of nanoparticles would be dissolved and recrystallized into microsphere assembly by nanoparticles under longer hydrothermal growth process. This result is consistent with the observation that surfactant assisted reaction is a simple and effective way to control the nucleation and growth process [35].

3.3.5.4 (b) Formation Mechanism of $LaFeO_3$ Microspheres Composite of Nanoparticles

The growth mechanism of microspheres can be determined by both the internal and external conditions such as molar concentration, temperature and pressure. It is well known that the shape of nanoparticles can be controlled by reactant and surfactants. The concentration of citric acid and La/Fe molar ratio play an important role in the formation of $LaFeO_3$ with sphere like morphology. The growth process of microsphere composed of $LaFeO_3$ nanoparticles can be proposed via the following reaction mechanism

$$La(NO_3)_3 \longrightarrow La^{3+} + 3NO_3^- \quad (1)$$

$$Fe(NO_3)_3 \longrightarrow Fe^{3+} + 3\,OH^- \longrightarrow Fe(OH)_3 \quad (2)$$

$$La^{3+} + C_6O_7H_8 \longrightarrow [C_6O_7H_5]^{3-}\,La^{3+} + 3H^+ \quad (3)$$

$$[C_6O_7H_5]^{3-}\,La^{3+} + 3OH^- \longrightarrow La(OH)_3 + [C_6O_7H_5]^{3-} \quad (4)$$

$$La(OH)_3 + Fe(OH)_3 \longrightarrow LaFeO_3 + 3\,H_2O \quad (5)$$

La^{3+} ions were produced when $La(NO_3)_3$ was dissolved in water (Eq. (1). $Fe(NO_3)_3$ ions dissociate into Fe^{3+} ions and further hydrolyze to form $Fe(OH)_3$ under hydrothermal reaction temperature as they are highly stable at room temperature (Eq.(2) [30]. La^{3+} combine easily with citric acid via the chelating effect than Fe^{3+} owing to the much higher stability constant than Fe^{3+} complex (Eq.(3). La^{3+} will be released slowly and gradually converted into $La(OH)_3$ via the alcoholysis reaction during hydrothermal process (Eq.(4). The simultaneously formed $Fe(OH)_3$ react with $La(OH)_3$ to form the $LaFeO_3$ precipitates by the condensation effect (Eq.(5). The surfactant and precursor plays a very important role on the final microsphere formation of $LaFeO_3$. It is known that citric acid not only accelerate the reaction speed of the growth units, but also directs their oriented-growth.

A plausible formation mechanism of $LaFeO_3$ microspheres is schematically illustrated in Figure 3.22. In the initial stage of hydrothermal reaction, metal ions in the solution react with citric acid to form relatively stable distorted primary particles of $LaFeO_3$ under supersaturated condition, which decreases the free metal ion concentration in solution and results in slow generation of primary particles. Generally, Ostwald ripening process involves in the formation of aggregates with primary crystallites through crystallization process due to the energy difference [35]. The formed $LaFeO_3$ nanoparticles aggregate together which is driven by the minimization of interfacial energy. The aggregated nanoparticles would act as primary cores that favour the homogenous assembly to produce larger self-ssembled aggregates via self-assembly process. The aggregated particles tend to form sphere shape, as the aggregation speed of nanoparticles from each direction to the core could be same. The effect of reaction time and calcination temperature on the $LaFeO_3$ microsphere morphology also supports the fact that Ostwald ripening effect is the essential mechanism.

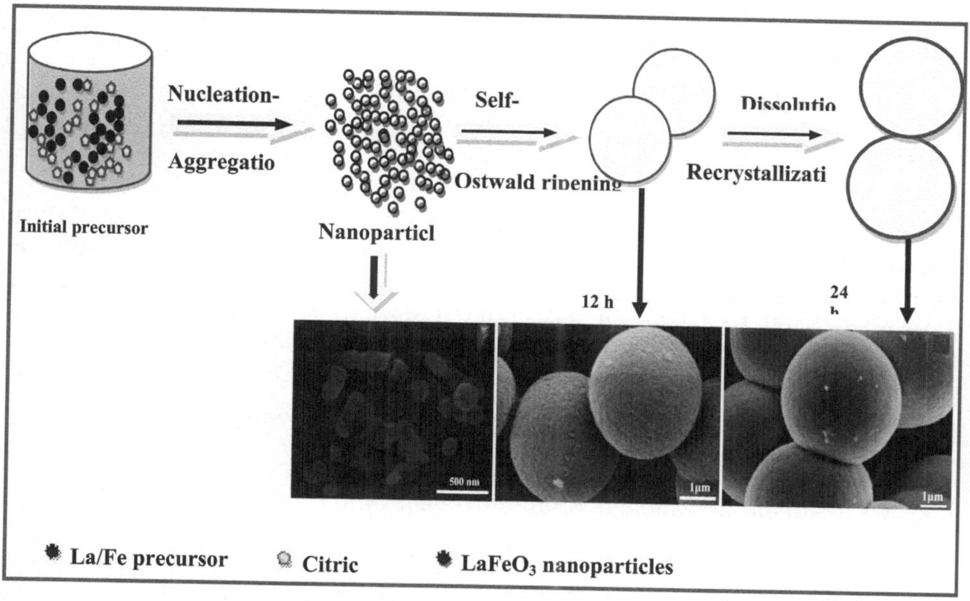

Figure 3.22. Schematic Representation of Formation Mechanism of $LaFeO_3$ Microspheres.

3.6 Conclusion

In summary, different morphologies of $LaFeO_3$ nanostructures such as nanocubes, nanorods nanospheres, dendrites nanostructures, flower like nanostructures and microsphere composed of nanoparticles have been successfully synthesized through facile and environmental friendly hydrothermal process without and in the presence of urea citric acid, CTAB and PVP as structure directing agents. Systematic investigation were performed to propose the possible growth mechanism of the obtained nanostructures. The hydrothermal conditions and surfactant concentration plays a major role in the formation of all $LaFeO_3$ nanostructures as investigated from morphological analysis. The structural and morphological changes were confirmed by XRD, Raman and SEM analysis. The average crystallite size of the different $LaFeO_3$ nanostructures were in the range 52 to 90 nm and the lattice fringes matches well with the crystallographic plane of all $LaFeO_3$ nanostructures which was also confirmed from TEM and HRTEM images.

The prepared materials were thermally stable and does not undergo any structural change even at higher temperature (~1000 °C) and have more adsorbed oxygen or oxygen in the hydroxyl group confirmed from XPS.

Finally, we arrive at a generalized conclusion that the following parameters play an important role in the morphology controlled synthesis of perovskite $LaFeO_3$ nanostructures

a. Preparation methods
b. Nature of precursors
c. Nature and concentration of surfactants
d. Growth conditions like reaction time, reaction temperature and calcination temperature.

Further, the morphology dependent physical properties, photocatalytic and biosensing applications of nanocubes, flow like nanostructures, dendrites nanostructures, nanorods, microspheres composed of nanoparticles and nanospheres will be discussed in the following chapters.

References

[1] P. Song, H. Qin, L. Zhang, K. An, Z. Lin, *et al., Sens. Actuat. B* **104** (2005) 312-316.

[2] T. Nakanishi, Y.Masuda, K. Koumoto, *Chem. Mater.* **16** (2004) 3484-3488.

[3] Y. Masuda and K. Kato, Cryst. *Growth Des.* **8** (2008) 2633- 2637.

[4] H. Rashid, T. K. Mandal, *J. Phys. Chem. C,* **111** (2007) 16750-16760.

[5] F.T. Li, Y. Liu, R. H. Liu, Z. M. Sun, D.S Zhao, *et al., Mater. Lett.* **64** (2010) 223-225.

[6] Fa-tang Li, Ying Liu, Zhi-min Sun, Rui-hong Liu, *et al., Mater. Lett.* **65** (2011) 406-408.

[7] L. Pei, M.Y. Li, J. Liu, B.F. Yu, J. Wang, *et al., Mater. Lett.* **64** (2010) 364-66.

[8] M. Sivakumar, A. Gedanken, W. Zhong, *et al., J. Mater. Chem.* **14** (2004) 764 -769.

[9] I. Coondoo, A. K. Jha, *Mater. Lett.* **63** (2009) 48-50.

[10] A.D. Jadhav, A. B. Gaikwad, V. Samuel, V. Ravi, *Mater. Lett.* **61** (2007) 2030-2032.

[11] Li X, Tang C, Ai M, Dong L, Xu Z, *Chem. Mater.* **22** (2010) 4879-4889.

[12] K. Holmberg, *J. Colloid Interfce Sci.,* **274** (2004) 355-364.

[13] Cao M, Liu T, Gao S, Sun G, Wu X, *Angew. Chem Int. Ed.* **44** (2005) 4197-4201.

[14] Y. Wang, G. Xu, L.Yang, Z. Ren, Z. *J. Cryst. Growth* **311**(2009) 2519-2523.

[15] Y. Wang, J. Zhu, L. Zhang, X. Yang, X. Wang, *et al., Mater. Lett.* **60** (2006) 1767-70.

[16] Z-X Wei, Y-Q Xu, H-Y Liu, *et al., J. Hazardous Mater.* **165** (2009) 1056 –1061.

[17] N. Uekawa, R. Yamashita, Y. J. Wu, *et al., Phys. Chem. Chem. Phy.* **6** (2004) 442 - 446.

[18] J. Ge, B. Tang, L. Zhuo, Z. Shi *et al., Nanotechnology* **17** (2006) 1316 – 1322

[19] H.Su, L. Jing, K. Shi, C. Yao, H. Fu, *J. Nanopart. Res.* **12** (2010) 967-974.

[20] Li Pengna, X. Hu, L. Zhang, D. Hongxing, L. Zhang, *Nanoscale.* **3** (2011) 974-946.

[21] M. Popa, J. Frantti, M. Kakihana, *Solid State Ionics.* **154 -155** (2002) 437- 445.

[22] Z-X Wei, Y-Q Xu, H-Y Liu, Hu *et al., J. Hazardous Mater.* **165** (2009) 1056 –1061.

[23] H. Su, L. Jing, K. Shi, C. Yao, H. Fu, *J. Nanopart. Res.* **12** (2010) 967-975.

[24] K. Rida, A. Benabbas, M. A.Pena, *et al.,* Appl. Catal. A **327** (2007) 173-179.

[25] S. Mickevicius, B. Vengalis, *et al., J. Alloys Compd.* **423** (2006) 107-111.

[26] X. Qi, J. Zhou, Z. Yue, Z. Gui, L. Longtu, *Mater. Chem. Phys.* **78** (2002) 25-29.

[27] Li FT, Y. Liu, R.H. Liu, Z.M. Sun, D.S. Zhao, C.G.Kou, *Mater. Lett.* **64** (2010) 223-25.

[28] H. G.Yang, H. C. Zeng, *J. Phys. Chem B.* **108** (2004) 3492-3495.

[29] S. Farhadi, Z.Momeni, M. Taherimehr, *J. Alloys Compd.* **471** (2009) L5-L8.

[30] He. Yaping, J. Zheng, Q. Sheng, *Analyst,* **137** (2012) 1031-1038.

[31] Y. Wang, G. Xu, L. Yang, Z. Ren, *J. Cryst. Growth* **311** (2009) 2519-2523.

[32] T. A. Witten, L. M. Sander, *Phys. Rev. Lett.,* **47** (1981)14001403.

[33] J. Ding, X. Lu, H. Shu, J. Xie, H. Zhang., *Mater. Sci. Eng. B* **171** (2010) 31-34.

[34] P. Sharma, J. Yeo, D. M. Kim, C. H. Cho, *J. Mater. Chem.* **22** (2012) 2838-2843.

[35] S. Farhadi, Z. Momeni, M. Taherimehr., *J. Alloys and Compd.* **471** (2009) L5.

Chapter 4

Morphology Dependent Physical Properties of $LaFeO_3$ Nanostructures

4.1 Introduction

The controlled synthesis of nanostructures with unique size and shape-dependent properties has been attracted increasing interest during the past decades [1,2]. Ordered nanostructures, assembled using nanoparticles, nanorods, nanospheres, nanosheets and nanoplates as building blocks have attracted great interest because of their novel and collective specific surface area, optical, magnetic and electrical properties in contrast to bulk materials [3]. The shape of the materials, which turn controlled by the preparatory methods govern the potentiality materials for specific applications. Hence in the present work, we made an attempt to study the morphology dependent properties of perovskite $LaFeO_3$ nanostructures. In the previous chapter we discussed the controlled growth and optimum experimental condition adopted in the growth of different morphologies of perovskite $LaFeO_3$ nanostructures. In this chapter, we will discuss the morphology dependent physical properties of different $LaFeO_3$ nanostructures like, nanocubes, microflowers/flower like, dendrites, nanorods, microspheres and nanospheres using specific surface area, optical, magnetic and electrical properties. The specific surface area and pore size distribution for the prepared morphology dependent $LaFeO_3$ nanostructures were measured from nitrogen adsorption-desorption isotherms using BET and BJH methods. The optical properties of different $LaFeO_3$ nanostructures were measured by UV-Vis diffuse reflectance spectroscopy. It will be found that the absorption cut-off wavelength varied depending on the morphologies. The magnetic properties are studied using VSM which shows variations in coercivity, remanent magnetization and saturation magnetization. The electrical behaviour is measured using impedance and electrical conductivity also shows disparity depending on the morphology.

4.2 Surface Properties of $LaFeO_3$ Nanostructures

4.2.1 Importance of Surface to Volume Ratio in Nanomaterials

The surface properties of nanoparticle relay on the drastic increase in surface to volume ratio which in turn paves way for many potential application [4]. Nanomaterials have a relatively larger surface area when compared to the same volume or mass of the material produced in a larger form. The increase in surface to volume ratio on decreasing the particle size down the nanoscale can be implicit by an layman example. Let us consider a cubic block of side 1 m, with surface area of 6 sq. m and a volume of 1 cubic meter. In this case, the surface to volume ration is 6. When the cubic block is cute into 8 equal piece of length 0.5 m, the surface area of each piece would be 1.5 sq. m. So, the total surface area of the 8 pieces would be 12. However, the volume remains same, but surface to volume ratio become 12. Further, the cubic block cute into 27 pieces, the surface to volume ratio is 18 [5]. Thus we find that when the given volume is divided to smaller pieces the surface area to volume ration increases several times. Hence as particle size decreases a greater proportion of atoms are found at the surface compared to core. Nanoparticles have a much greater surface area per given volume compared with larger particles. It makes nanomaterials are more chemically reactive. A significant number of technological advantages of nanomaterials develop this property for surface related applications. The requirement of high surface to volume ratio becomes very important, when the process is surface dependent and the surface is the rate limiting step for functional applications.

4.2.2 Significance of Porosity

Further significant surface related properties of nanomaterials exhibit with high surface to volume ratio are the porosity. Porosity refers to the volume of pores in a solid. Pores on the surface of a materials, vary with shape and size, which contribute to the internal surface of the materials. Highly porous solid materials, particularly those exhibits well-defined pore sizes, have been exploited in a wide range of applications, such as gas sensor, biosensor, catalytisis and so on. According to IUPAC standards, porous materials can be classified according to their size as [6]

(a) Macrospores (pores with > 50 nm)

(b) Mesopores (pores with 2-50 nm)

(c) Microspores (pores with < 2nm)

In the case of gas sensor and biosensor applications, porosity offers adsorption sites for the incoming molecules (gas) to adsorb and interact with the semiconductors and desorb depending on the nature and extent of pores on the surface. For catalytic applications. Porosity in the surface enhances the separation and transfer of photo induced electron hole pair, which otherwise would recombine, and limit the photocatalytic efficiency.

4.2.3 (a) Brunauer, Emmett and Teller (BET) Methods – Specific Surface Area

Adsorption measurements have a variety of uses and applications. For example; powder materials, sinter frits, filters, ceramics, wash coats, catalysts, polymers, and coat tape *etc.* The most important one is determining surface area of perovskites. The method of measuring total surface area of perovskite is by adsorption of particular molecular species from gas or liquid onto the surface. If the condition under which a complete adsorption layer, averaging one molecule thick, can be established and the area covered per molecule is known, then the quantity of adsorbed material gives directly the total surface area of the perovskite. The most common method of measuring surface area is calculated by Brunauer, Emmett and Teller (BET) methods [7].

When a highly dispersed solid is exposed in a closed space to a gas or vapor at some definite pressure, the solid begins to adsorb the gas. Adsorption on porous materials proceeds via monolayer adsorption followed by multilayer adsorption and capillary condensation [7]. At a low relative pressure, a monolayer of adsorbent is absorbed on the surface like non porous materials. The amount of adsorbent is used to calculate the surface area. Using the Brunauer, Emmett and Teller (BET) theory, one can calculate the specific surface area of a solid sample from the number of monolayer gas molecules required to cover the solid surface, and the cross-sectional area of the gas molecule being adsorbed. The following linear relationship exists between the volume adsorbed, v (cm^3/g) and the relative vapour pressure p/p_0. The BET equation is written as:

$$\frac{1}{W\left(\left(P_0/P\right)-1\right)} = \frac{1}{W_m C} + \frac{C-1}{W_m C}\left(\frac{P}{P_0}\right)$$

Here, W is the weight of gas adsorbed, P/P_0 is the relative pressure, W_m is the weight of adsorbate as monolayer, C is the BET constant. The values of W_m and c are obtained from the slope of the plot. The surface area is then calculated from $\tilde{o}_m$ and the area occupied by one molecule of the adsorbte. The segment in the application of the BET method is the calculation of the surface area (often termed the BET area) from the monolayer capacity. This requires the average area a_m (molecular cross section area), occupied by the adsorbate molecule in the complete monolayer and value of W_m obtained from the slope of the linear plot.

$$A\ (BET) = W_m^{a} .\ L .\ a_m$$

From the total surface area A (BET), the specific surface area a(BET) can be calculated from the equation

$$A\,(BET) = \frac{A\,(BET)}{m}$$

Where, m is the mass of the adsorbent and L is the Avogadro constant.

4.2.4 (a) Adsorption – Desorption Isotherms

The bonding in physisorbed system is weak, Vander Waals interaction and it tends to be reversible, in the sense that the surface layer is always in equilibrium with the molecules of the gas phase. The textural characterization, such as surface area, pore volume and the pore size distribution of the as-prepared aerogels and oxides were obtained by N_2 physisorption at

77 K using a Micromeritics ASAP 2010. Prior to N_2 physisorption, the samples were degassed at 200 °C under vacuum for 2 hours. This N2 adsorption isotherms, can be used to calculated, the specific surface area was calculated. The mesoporous volume (VBJH), the average pore diameters and the pore size distributions were estimated by the BJH method applied to the desorption branch of the isotherm. This theoretical information related to the surface area and pore volume calculations were used in the present work.

(b) Characteristics of Physical Adsorption

- ☆ Low heats of adsorption, no violent or disruptive structural changes.
- ☆ Can involve multiple layers of adsorbate, thus allowing for pore measurements.
- ☆ High temperatures tend to inhibit physical adsorption.
- ☆ Adsorption equilibrium is achieved quickly since no activation energy is generally required.
- ☆ Physical adsorption is fully reversible, allowing adsorbate to fully adsorb and desorbte.

Here, we discuss the relation for constant temperature between the quantity of gas adsorbed and the pressure with which it is in equilibrium is termed an adsorption isotherm. The physical adsorption isotherms may be grouped into six types, which are frequently referred to as the Brunauer, Deming, Deming and Teller (BDDT) classification as in Figure 4.1. The Type I curves represents the typically micropores with the exposed surface residing almost exclusively inside the micropores, which once filled with adsorbate and leave little or no external surface for further adsorption. The Type II curve most frequently found when adsorption occurs on nonporous powders or powders with diameters exceeding micropores. Inflection point occurs near the completion of the first adsorbed monolayer. A Type III is characterised by heats of adsorption less than the adsorbate heat of liquification, adsorption proceeds as the adsorbate interaction with an adsorbed layer is greater than the interaction with the adsorbent surface. The Type IV Occur on mesoporous capillary condensation i.e porous adsorbents with pores in the range of 1.5 - 100 nm. At higher pressures the slope shows increased uptake of adsorbate as pores become filled, inflection point typically occurs near completion of the first monolayer. The Type V are observed where there is small adsorbate-absorbent interaction potentials (similar to type III), and are also associated with pores in the 1.5 - 100 nm range. Type VI isotherm is associated with a stepwise multilayer adsorption on a uniform non-porous surface (layering).

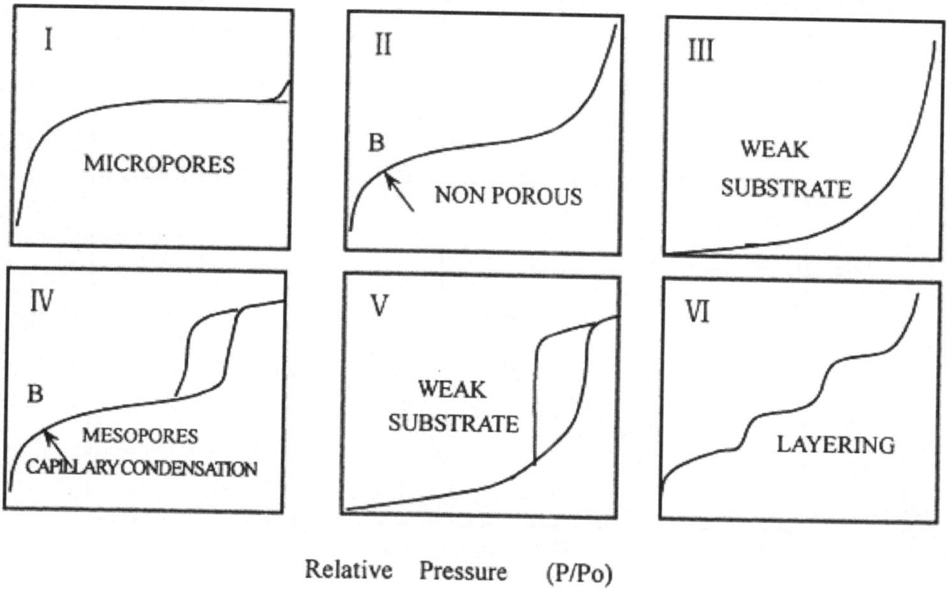

Figure 4.1. Different Types of Physisorption Isotherm.

Hysteresis appearing in isotherm (as in type IV) is usually associated with capillary condensation in mesoporous structures. The capillary condensation is used for the determination of pore size, dependent on the shape of the hysteresis loop as show in Figure 4.2.

The H1 type of hysteresis loop has a vertical and nearly parallel adsorption and desorption, over an appreciable range of gas uptake.On contrary, the H4 type of hysteresis is characterized by a horizontal and parallel isotherm. The H2 and H3 type of hysteresis may be as intermediate between H1 and H4. Although the effect of various factors on adsorption hysteresis is not fully understood. The shape of hysteresis loops can be related to specific pore structures. Type H1 is often associated with agglomerates or compacts of approximately uniform spheres in fairly regular array. Many porous adsorbents (inorganic oxide gels) tend to give Type H2 loops. But in such systems the distribution of pore size and shape is not well defined. Indeed, the H2 loop is especially difficult to interpret. Type H3 loop, which does not exhibit any limiting adsorption at high P/P_0, is observed with the aggregation of plate like particles giving rise to slit-shape pores. Similarly, the Type H4 loop is often associated with narrow slit like pores.

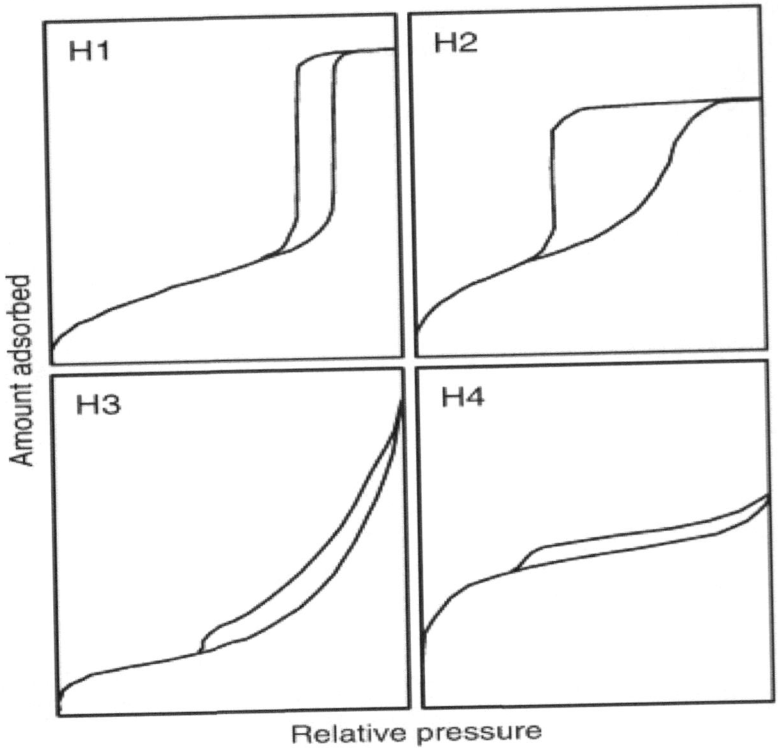

Figure 4.2. Different Shapes of Hysteresis Loop.

4.2.5 Specific Surface Area and Pore Size Distribution of $LaFeO_3$ Nanostructures

The surface properties of nanostructures materials are very important to achieve a good sensing and catalytic performance. In the present work, surface area of morphological different $LaFeO_3$ nanostructures were measured from nitrogen adsorption - desorption isotherms using Brunauer-Emmett-Teller (BET) and Barrett–Joymer–Halenda (BJH) method. In general, the specific surface area is defined as the ratio between the absolute surface area of a solid and mass. The BET method involves the determination of the amount of adsorbate gas required to cover the external and the accessible internal pore surfaces of a solid with a complete monolayer of adsorbate. The adsorption isotherm is thus a plot of the amount of gas adsorbed (in mol g^{-1}) as a function of the relative pressure [8]. Nitrogen adsorption and desorption isotherms of the different $LaFeO_3$ nanostructures with their corresponding pore size distribution evaluated using BJH method are shown in Figures 4.3-4.6. The inset in each figure maps the corresponding pore size distribution calculated using BJH method. The specific surface area and pore size distribution of the different morphologies prepared in the present work determined by using BET and BJH method is tabulated in Table 4.1.

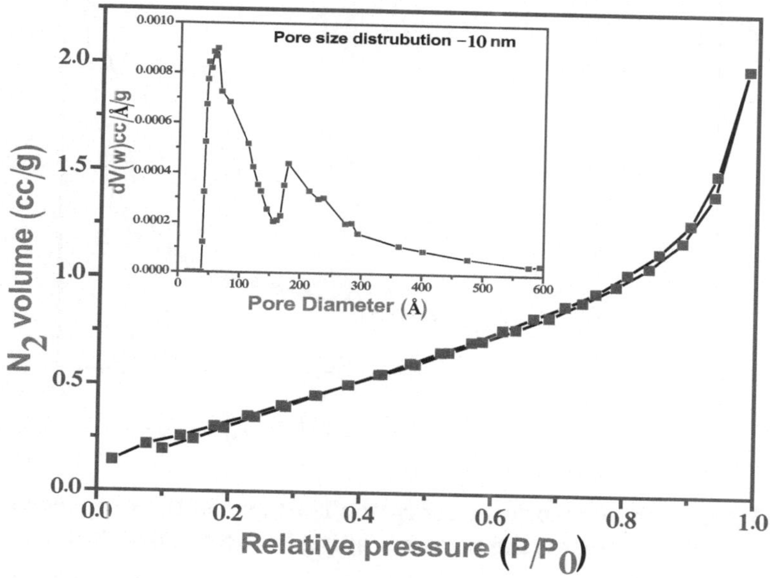

Figure 4.3. Nitrogen Adsorption-Desorption Isotherm of the Prepared $LaFeO_3$ Nanocubes and (Inset) the Corresponding Pore Size Distribution.

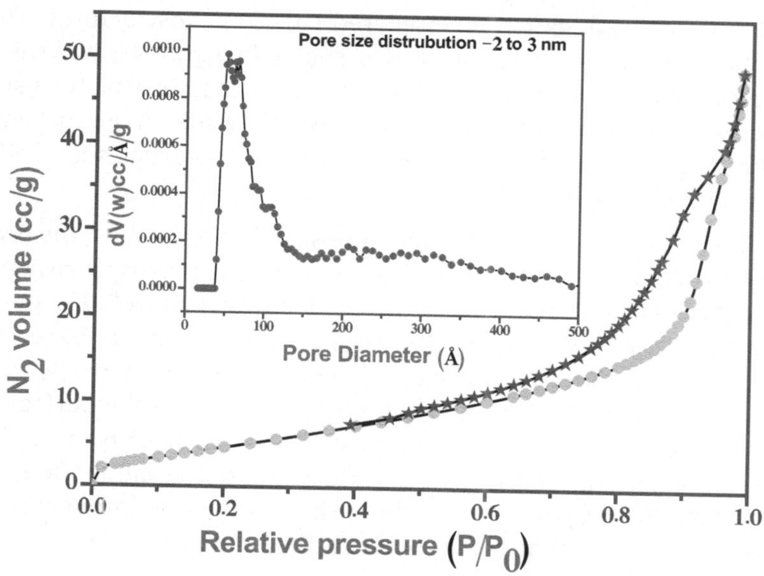

Figure 4.4. Nitrogen Adsorption-Desorption Isotherm of the Prepared $LaFeO_3$ Microspheres Composed of Nanoparticles and (Inset) the Corresponding Pore Size Distribution.

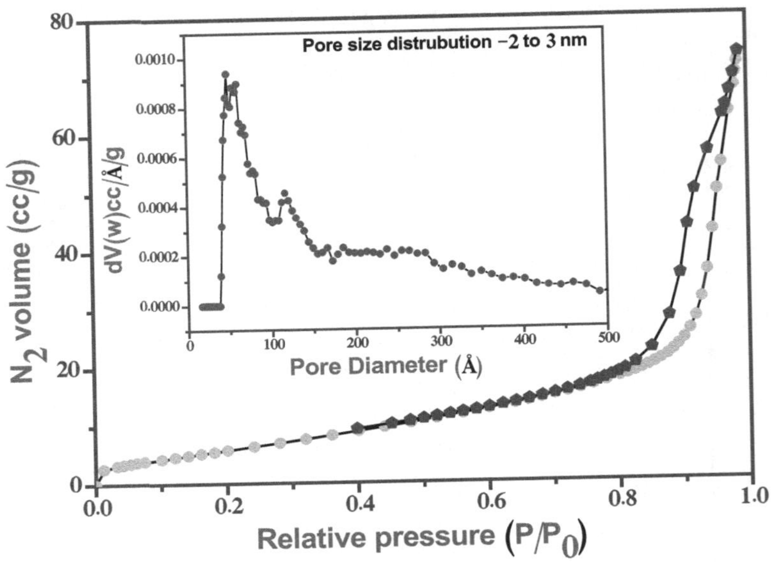

Figure 4.5. Nitrogen Adsorption-Desorption Isotherm of the Prepared $LaFeO_3$ Nanospheres (Inset) the Corresponding Pore Size Distribution.

The hysteresis loop in the high-pressure range ($0.8 < P/P0 < 1$) can probably be associated with the larger pores that could be formed between secondary particles due to the aggregation for prepared $LaFeO_3$ nanostructures. The BET specific surface area of $LaFeO_3$ nanocubes as shown in Figure 4.3 was calculated to be 40.7 m^2g^{-1}, the pore size distribution of the nanocubes calculated using BJH methods represents the presence of measopores o the 6 nm on the surface of the sample. The shape of the hysteresis exhibited for these particles is of H3 type, characterized by slit shaped pores.

In addition, specific surface area and pore size distribution for the prepared different morphologies of $LaFeO_3$ nanostructures such as microsphere composed of nanoparticles and nanospheres are shown in Figure 4.4 and 4.5. The hysteresis loop in the high-pressure range ($0.8 < P/P0 < 1$) can probably be associated with the larger pores that could be formed between secondary particles due to aggregation for prepared $LaFeO_3$ nanostructures. The BET specific surface area of $LaFeO_3$ microspheres and nanospheres were calculated to be 45.8 and 48.9 m^2g^{-1}, respectively. In addition, the BJH method was used to determine the pore size distribution curve (inset) and peak position shows the mesoporous nature of 2 to 3 nm. From the shape of the hysteresis loop, they were found to be of H3 type, which is again characterized by slit shape pores.

Figure 4.6 and 4.7 shows the isotherm recorded for the flower like nanostructures and dendrites like nanostructures. The distinct hysteresis loop from 0.8 to 1 p/P0 in the case of flower and dendritic nanostructures respectively can be categorized as Type IV isotherm. The flower like nanostructures had a specific surface area of 63.2 m^2g^{-1} and narrow size mesoporous distribution of 1 to 10 nm. The dendrites are 68.5 m^2g^{-1} and broad measopore distribution of 1 to 10 nm.

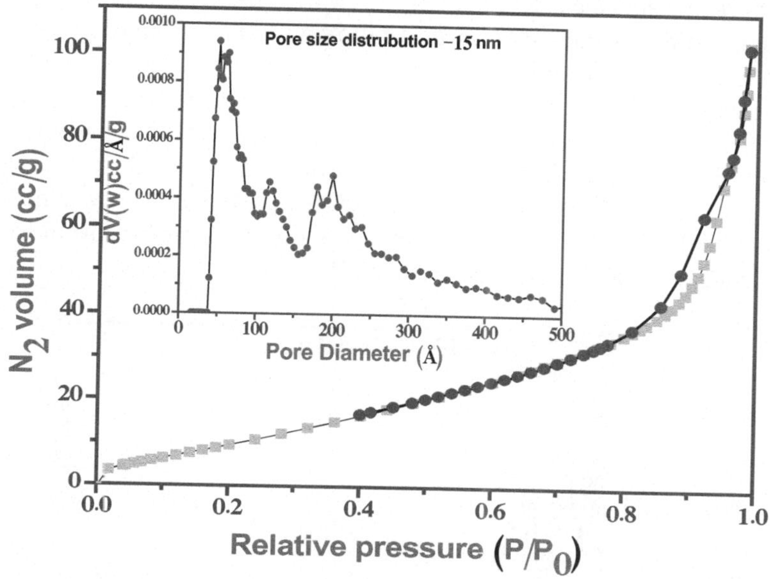

Figure 4.6. Nitrogen Adsorption-Desorption Isotherm of the Prepared $LaFeO_3$ Flower like Nanostructures and (Inset) the Corresponding Pore Size Distribution.

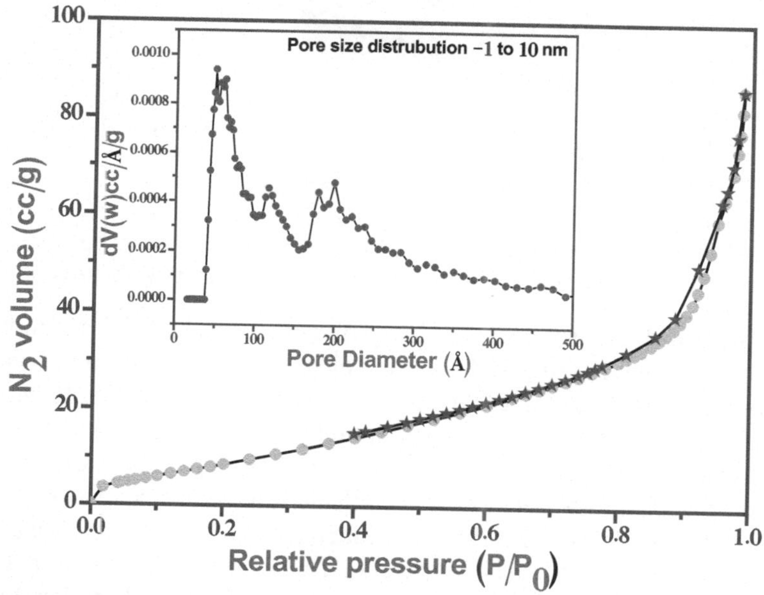

Figure 4.7. Nitrogen Adsorption-Desorption Isotherm of the Prepared $LaFeO_3$ Dendrites Nanostructures (Inset) the Corresponding Pore Size Distribution.

In the both the case the shape of hysteresis loop, found to be of H3 type, which is again characterized by slit shaped pores. The pores can be ascribed to the inter-particle space and the inter branch space. Moreover, the whole architecture is relatively large, the surface area can likely be contributed to the surface condition of flower and dendrites nanostructures. In the pore size distribution curve, both structures, and three different peaks can be clearly noticed. The first and second peaks may correspond to the voids between the crystallites present in agglomerated particles. The presence of agglomerated particles in the $LaFeO_3$ material with varying size has been confirmed earlier in this study from the obtained SEM micrographs. The presence of crystallites within the aggregated particles was evidenced from the XRD results. Figure 4. 8 Shows the nitrogen adsorption-desorption isotherm of the nanorods of Type IV with hysteresis of H3 type.

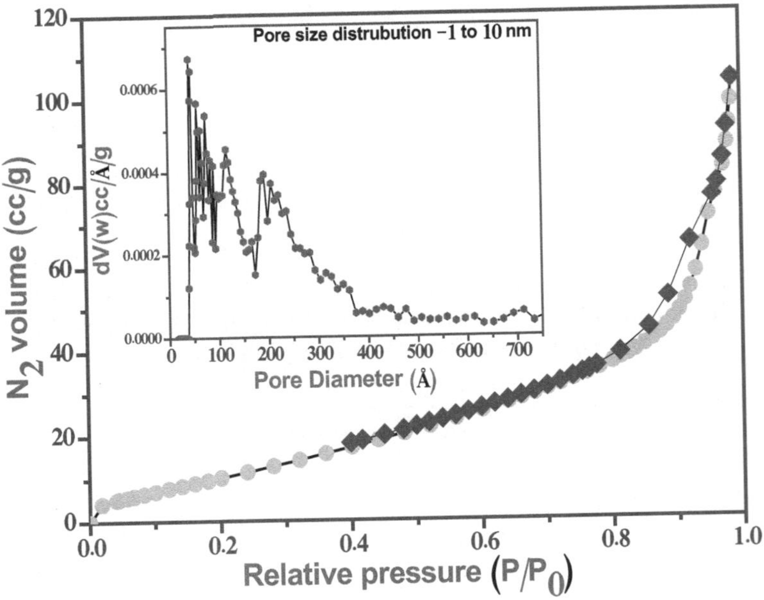

Figure 4.8. Nitrogen Adsorption-Desorption Isotherm of the Prepared $LaFeO_3$ Nanorods (Inset) the Corresponding Pore Size Distribution.

Of all the nanostructures prepared in the present work, $LaFeO_3$ nanorods have a very high specific surface area of 84.4 m^2/g. The pore size distribution curve shows that the dominant peaks are in the mesoporous range of 1 to 10 nm. It can also be observed that the pores are bimodal, signified by the presence of small and large mesoporous on the surface of the samples. Among all the morphologies investigated in the present work, nanocubes shows relatively lower surface area because the prepared without surfactant. All the other morphologies were prepared using the different surfactant such as, urea, citric acid, PVP and CTAB. The nanostructure prepared with the presence of urea having the shape of nanorods and it shows the high surface area because favoured of certain reaction intermediates. The topotactic transformation of this intermediate desorption of surface adsorbed CO_3 originated during the decomposition of urea, played a major role in enhancing the surface area and pore size distribution in this sample. The role of urea and the intermediates formed at various stages of the reaction has already discussed in Chapter 3.3.5. The above results confirm, the prepared all $LaFeO_3$ nanostructures could show high biosensing property and enhanced photocatalytic performance as the properties are closely related to the morphology, surface area and pore size distribution.

Table 4.1. The Specific Surface Area and Pore Size Distribution of the Prepared different $LaFeO_3$ Morphologies

Nanostructures	*Specific Surface Area (m^2/g)*	*Pore Size Distribution (nm)*	*Type*
Nanocubes	40.7	6	Type IV isotherm H3 hysteresis
Microspheres	45.9	2 to 3	Type IV isotherm H1 hysteresis
Nanospheres	48.9	2 to 3	Type IV isotherm H1 hysteresis
Flowerlike nanostructures	63.2	1 to 10	Type IV isotherm H3 hysteresis
Dendrites nanostructures	68.5	15	Type IV isotherm H3 hysteresis
Nanorods	84. 4	1 to 10	Type IV isotherm H3 hysteresis

4.3 Optical Properties of different $LaFeO_3$ Nanostructures

Compounds with the perovskite structure have been used as model systems for spectroscopic studies in the infrared, visible, and ultraviolet region. Perovskite oxides (ABO_3) are particularly suited since the effect of magnetic ordering of the transition metal ion in the B-site can be studied in addition to that of the rare earth ion A. Because the difference in ordering temperature of the two types of ions is large there is little interaction. The absorption spectra either for the host ion of the perovskite structure or after incorporation of a low concentration "probe" into a diamagnetic lattice can be studied. In the former case of magnetically concentrated systems, a broad spectrum is obtained, with line widths typically of 0.5-5 cm^{-1}, while in the latter a sharp spectrum is obtained, with line widths of about 0.1 cm^{-1}. The optical properties of many ABO_3 compounds, in which B is a transition metal ion with an empty or partially filled d-shell, have been explored. For the $LaFeO_3$ perovskite containing an empty d-shell transition ion (Fe^{3+}), the emission spectrum shows luminescence which is independent of sample history and contamination and shows a band maximum at about 550 nm, with low quenching temperature (35 K) and a small Stokes shift (7000 cm-1). The Fe octahedral sharing corners in $LaFeO_3$

are favorable for energy band formation. The broadening of the energy levels of the Fe octahedron results in a considerably lower energy for the first absorption transition, and the excited state of the Fe becomes mobile. This situation has been described due to the coexistence of free and delocalized (self-trapped) excitons [9].

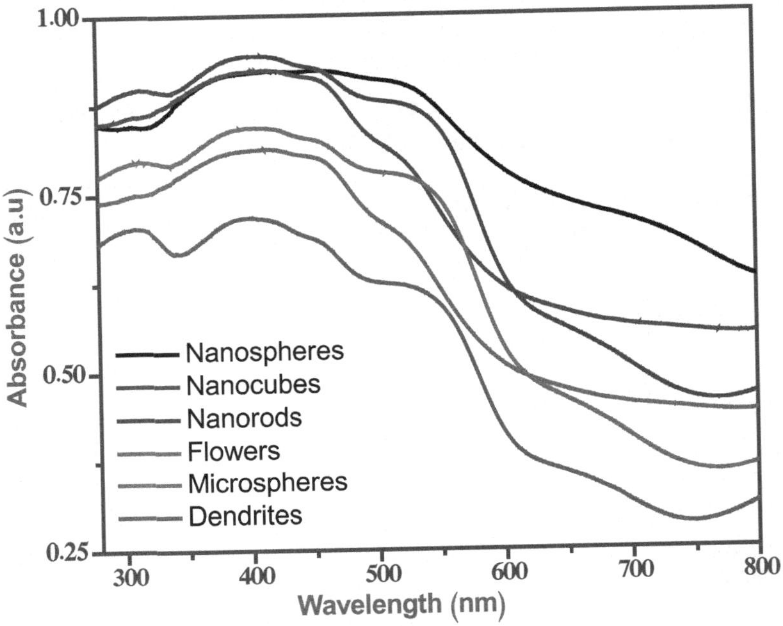

Figure 4.9. UV-Vis DRS of different $LaFeO_3$ Nanostructures.

A large body of work on the optical properties of metal ions with a partially filled d-shell containing perovskites exists. Ions with the d^3 configuration have been extensively used to investigate the optical properties of perovskites. The optical transitions involved were shown to be of the metal to- metal charge-transfer type, *e.g.*, Fe^{3+}, Cr^{3+} et., which are responsible for the brownish colour of the composition. The identification of the nature and structure of the exposed sites on the perovskite surfaces as well as the possible involvement of these sites in catalytic reactions have frequently been studied by the adsorption of suitable probe molecules. Both the equilibrium and the kinetics of adsorption and, in some cases, the desorption of molecules from the surface were analysed. Optical properties of the perovskite oxide materials were investigated by UV-Vis diffuse reflectance spectroscopy.

The energy band features of perovskite oxide materials are considered the key factor in determining its photocatalytic activity. It can be observed from Figure 4.9, that the six different morphologies of $LaFeO_3$ nanostructures have similar and strong visible light absorption ability. The absorption cut-off wavelength in the range 580 to 620 nm indicates that the catalyst exhibits a typical optical absorption of perovskite oxide from UV to 620 nm. In perovskite-type oxides the strong absorption edge can principally be attributed to the electronic transition from the valance band to conduction band, O 2p Fe 3d. The diffuse reflectance spectrum of $LaFeO_3$ allows us to calculate the optical band gap based on the absorption edge using the formula, $E_g = 1240/\lambda_g$ [10]. From the absorption measurements, the onset absorption wavelength and the corresponding bandgap energies of $LaFeO_3$ nanostructures such as nanocubes, flowers, dendrites, nanorods, microspheres and nanospheres are 2 - 2.27 eV, respectively as shown in Table 4.2. It has been reported in literature that the photocatalytic activity of catalysts depended on many factors, such as crystallinity, surface properties, absorption and optical properties. Under visible light irradiation, the photocatalyst absorbs light above 400 nm, which helps in the formation of photogenerated electron–hole pairs in semiconductors (11). Therefore, $LaFeO_3$ nanostructures posses suitable band gap that can be activated by visible light for photocatalytic decomposition of organic contaminants.

Table 4.2. Optical Parameters of different $LaFeO_3$ Nanostructures

Nanostructures	*Absorption Cut-off Wavelength (nm)*	*Bandgap Energy (eV)*
Nanocubes	582	2.14
Flowers	591	2.10
Dendrites	601	2.07
Nanorods	609	2.05
Microspheres	614	2.03
Nanospheres	619	2.00

4.4 Magnetic Properties of $LaFeO_3$ Nanostructures

4.4.1 Magnetic Materials

$LaFeO_3$ is a member of rare-earth orthoferrites ($RFeO_3$) which are a family of canted antiferromagnets with unusual variety of magnetic properties and their magnetic behavior had been extensively studied. Magnetic behaviour is one of the important and fundamental properties of $LaFeO_3$. As discussed in Chapter 1, now it is interesting and important to study the magnetic properties of $LaFeO_3$ not only for scientific interests but also for applications. In this context, the magnetic properties of $LaFeO_3$ have been studied extensively. But, the magnetic study of $LaFeO_3$ nanoparticles is rare. Antiferromagnetic nanoparticles always show unusual magnetic properties due to the finite-size effects, surface anisotropy effects shape anisotropy effect and so on [12-15]. So, it is interesting to study the magnetic properties of $LaFeO_3$ nanoparticles. However, due to reduction of particle size to nanoscale, the magnetic moments slightly deviate from the exact antiparallel alignment. This is known as spin canting.

In our present work to analyze size and shape dependent properties of the different $LaFeO_3$ nanostructures. The most common measurement method employed for studying the magnetic behaviour of a material is by determining the hysteresis loop using a Vibrating Sample Magnetometer (VSM).It measures the magnetization (M) of a material as a function of applied magnetic field (H) and hence, it is often referred to as M-H loop.

4.4.2 M-H Hysterias Loop

A great deal of information can be learned about the magnetic properties of a material by studying its hysteresis loop [13]. A hysteresis loop shows the relationship between the induced magnetic flux density **(B)** and the magnetizing force **(H).** It is often referred to as the B-H loop. An example hysteresis loop is shown in Figure 4.12.

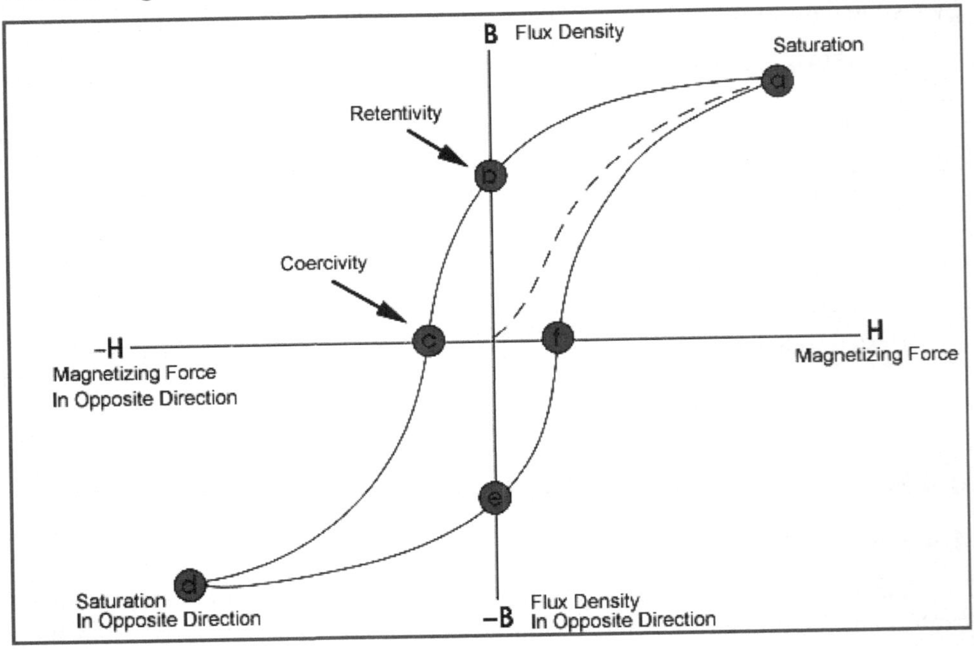

Figure 4.12. Schematic Illustration of Magnetic Hysteresis Loop.

The loop is generated by measuring the magnetic flux of a ferromagnetic material while the magnetizing force is changed. A ferromagnetic material that has never been previously magnetized or has been thoroughly demagnetized will follow the dashed line as **H** is increased. As the line demonstrates, the greater the amount of current applied **(H+)**, the stronger the magnetic field in the component **(B+)**. At point "a" almost all of the magnetic domains are aligned and an additional increase in the magnetizing force will produce very little increase in magnetic flux.

The material has reached the point of **magnetic saturation**. When His reduced to zero, the curve will move from point "a" to point "b." At this point, it can be seen that some magnetic flux remains in the material even though the magnetizing force is zero. This is referred to as the point of retentivity on the graph and indicates the **remanence** or level of residual magnetism in the material. (Some of the magnetic domains remain aligned but some have lost their alignment.) As the magnetizing force is reversed, the curve moves to point "c", where the flux has been reduced to zero. This is called the point of **coercivity** on the curve. (The reversed magnetizing force has flipped enough of the domains so that the net flux within the material is zero.) The force required to remove the residual magnetism from the material is called the coercive force or coercivity of the material.

As the magnetizing force is increased in the negative direction, the material will again become magnetically saturated but in the opposite direction (point "d"). Reducing **H** to zero brings the curve to point "e." It will have a level of residual magnetism equal to that achieved in the other direction. Increasing **H** back in the positive direction will return **B** to zero. Notice that the curve did not return to the origin of the graph because some force is required to remove the residual magnetism. The curve will take a different path from point "f" back to the saturation point where it complete the loop.

From the hysteresis loop, a number of primary magnetic properties of a material can be determined. But three major magnetic parameter that can be determined from the hysteresis loop are

Saturation magnetization (Ms): The maximum magnetization of a ferromagnetic material, when all the magnetic moment of the samples is oriented along the applied field direction.

Remenant magnetization (Mr): It is a measure of the magnetization retained by the sample when the external applied field is brought to zero.

Coercive Force: It is a measure of reverse magnetic field which must be applied to a magnetic material to make the magnetization to Zero

4.4.3 Shape of Hysteresis Loop

The shape of the hysteresis loop tells a great deal about the material being magnetized. The hysteresis curves of two different materials are shown in the graph.

Relative to other materials, a material with a wider hysteresis loop has:

- ☆ Lower Permeability
- ☆ Higher Retentivity
- ☆ Higher Coercivity
- ☆ Higher Reluctance
- ☆ Higher Residual Magnetism

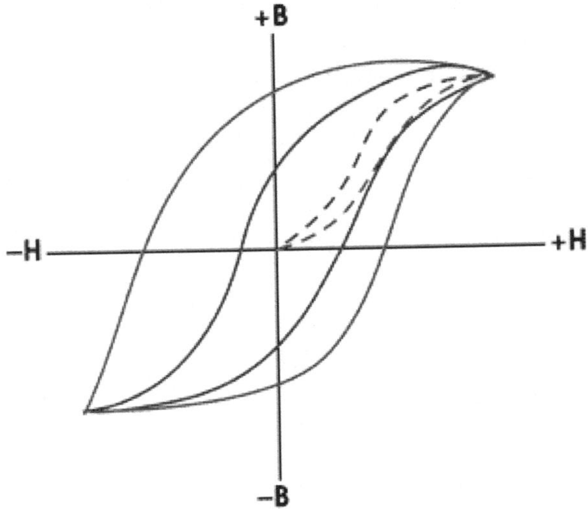

Relative to other materials, a material with the narrower hysteresis loop has:

- ☆ Higher Permeability
- ☆ Lower Retentivity
- ☆ Lower Coercivity
- ☆ Lower Reluctance

In magnetic particle testing, the level of residual magnetism is important. Residual magnetic fields are affected by the permeability, which can be related to the carbon content and alloying of the material. A component with high carbon content will have low permeability and will retain more magnetic flux than a material with low carbon content.

The shape of the M-H loop and the most important magnetic parameters, which determine the applicability of a magnetic material in various applications, dependent mainly on two factors,

- ☆ Magnetic Anisotropy
- ☆ Type of magnetic domain

4.4.3.1 Magnetic Anisotropy

The dependence of magnetic properties on a preferred direction (z-axis) is called magnetic anisotropic. The magnetic anisotropy is generally associated with ferromagnetic materials. The anisotropy energy creates barriers to free rotation of the magnetization within the magnetic crystal. This leads to different magnetic behaviour of a same materials with different size, shape and crystal structure. There are many causes for anisotropy energy. The most important ones are

1. Magneto crystalline anisotropy- crystal structure
2. Shape anisotropy – grain shape

Magnetocrystalline Anisotropy

Magnetocrystalline anisotropy is an intrinsic property which depends on the crystal orientation in a sample [16]. It depend on the crystallographic orientation of the sample in the magnetic field, the magnetization reaches saturation in different field. Magnetocrystalline anisotropy energy is the energy necessary to deflect the magnetic moment in a single crystal from the easy to the hard direction. When magnetized and demagnetized along the easy axis, coercitivty value is lower. On contrary, the hysteresis loop shows a high coercivity value which when magnetized along the hard axis.

Shape Anisotropy - Grain Shape

Shape of a material/grain is another source of magnetic anisotropy in ferromagnetic materials. For example, consider a long thin needle shaped grain. The anisotropy energy (Ea) will be less if the magnetization is along the long axis than along one of the short axes. This produces an easy axis of magnetization along the long axis. A sphere, on the other hand, has to shape anisotropy [17]. Enhanced anisotropy induces large magnetic coercivity, where the magnetic spins are preferentially aligned along the long axis and their reversal in the opposite direction requires higher energies than that for spheres.

4.4.4 Type of Magnetic Domain

A remarkable property of ferrimagnetic materials is not so much that they have a spontaneous magnetization, but rather that their magnetization can be influenced by the application of very low magnetic fields. Even the earth's field can cause magnetization changes even though the interatomic exchange forces responsible for the spontaneous magnetization are equivalent to a field of about 1000 T, almost 100 million times greater than the earth's field. What allows this to occur is the fact that the sample is actually composed of small regions called magnetic domains, within each of which the local magnetization is saturated but not necessarily parallel. Weiss's theory, a ferromagnetic sample is composed of small regions called magnetic domains, within each of which the local magnetization is saturated is saturated but not necessarily parallel. Domains are small (1-100's microns), but much larger than atomic distances.

On the other hand if the dimension of the material is reduced drastically, until a few nanomaterials, these domain walls will be forced to shrink. This proximity (density) of domains when the material reaches the dimensions of nanometre scale, yields the domains to merge into a single and creating what is known as a single domain particle. The exact dimension at which a nanostructure materials passes from single - to - multi domain depends on each material and is known as critical size. Thus the domain states in different magnetic minerals will have different grain size dependence. However, as a rule of thumb, the larger the grain size the more domains it contains.

The existence of domains is hinted at by the observation that some magnetic properties, and in particular, coercivity and remanence vary greatly with grain size. The magnetic behavior can be subdivided on the basis of grain size into four ranges

- ☆ **SPM: superparamagnetic**
- ☆ **SD: single domain**
- ☆ **PSD: pseudo-single domain**
- ☆ **MD: multidomain**

The maximum coercivity for a given material occurs within its SD range. For larger grain sizes, coercivity decreases as the grain subdivides into domains. For smaller grain sizes, coercivity again decreases, but this time due to the randomizing effects of thermal energy. Domains constitute a fundamental concept in magnetism. A ferro or ferrimagnetic material may be generally defined as one that possesses a spontaneous magnetization (Ms) dependent on temperature, but only slightly dependent on applied field.

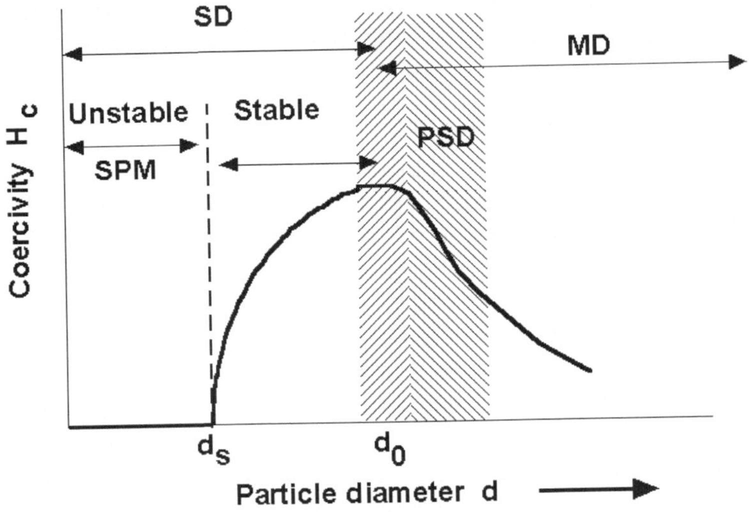

Figure 4.13. Variation of Intrinsic Coercivity with Particle Size.

As the grain size decrease, a critical size will be reached where the grain can no longer accommodate a wall. Below this critical size, the grain contains a single domain (SD). An SD grain is uniformly magnetized to its saturations magnetization. Changing the magnetization of a MD grain is an energetically easy process, which needs translation of the domain wall. Hence it can be accomplished in relatively low fields; Thus MD grains are magnetically soft with low values of coercivities and remancence. However, the only way to change the magnetization of a SD grain is to rotate the magnetization, an energetically difficult process. Thus, SD grains are magnetically hard and have high coercivities and remanence. The critical size for SD behaviour dependent on several factors including, the saturation magnetization and the shape of the grain.

The distinction between SD and MD is straightforward. However, small MD grains exhibit a mixture of SD like (high remanence) and MD like (low coercivity) behaviour. This state is called pseudo single domain (PSD), which occurs in smaller particles with few multi-domain grains. A particle size continues to decrease with the SD range, another critical threshold is reached, at which remanence and coercivity go to zero. When this happens, the grain becomes superparamagnetic (SPM). An SD particle approaching the SPM size has a uniform magnetization directed along the easy axis of magnetization. If the volume is small enough, the thermal energy will be sufficient to overcome the anisotropy energy separating the magnetizations states and cause a spontaneous reveal of magnetization. Figure 4.13 shows the variation of coercivity with particle diameter (equivalently particle size) and the corresponding domain type demarcation as SD, MD, SPM and PSD.

The coercivity of small particle has a striking dependence on their size, as the particle size reduced. It is typically found that the coercivity increases goes through a maximum and then tends towards zero. The maximum coercivity for a given material occurs within its PSD range, which is characterized by the presence of multidomains in smaller particles. For larger grain sizes, coercivity decreases as the grain subdivides into domains. For grain sizes lesser than the PSD range, the particles are purely SD. For very small grain sizes, coercivity again decreases, because of the randomizing effect of thermal energy as seen in superparamagnetisim. The PSD range thus marks an upper limit to the SD range and lower limit to the MD range.

4.4.5 Magnetic Properties of different Morphologies $LaFeO_3$ Nanostructures

The magnetic behaviour of the different morphologies different of $LaFeO_3$ nanostructures prepared in the present work have been studied by measuring the magnetic hysteresis using VSM at room temperatures maximum applied magnetic field 20 KOe. $LaFeO_3$ is known to be antiferromagnetic with weak ferromagnetism [18]. The net magnetization of $LaFeO_3$ should be very small because of the antiferromagnetic order of the Fe^{3+} spins. Only a slight canting of the adjacent Fe^{3+} spins produced weak ferromagnetism. The spontaneous magnetization of $LaFeO_3$ bulk crystals was only ~0.1 emu/g at room temperature [18]. The shape of hysteresis loops showing the small spontaneous magnetization were characteristics of weak ferromagnetism. All the morphologies exhibit slight hysteresis features, suggesting the weak ferromangnetic behaviour at room temperature. The M-H loop of the six different morphologies of $LaFeO_3$ are shown in Figure 4.14, The hysteresis loops doesn't saturation even up to the maximum applied magnetic field of 20 T. It indicates the presence of large anisotropy and uncompensated surface spins, *i.e* the presence of more surface atoms in the nanomaterials.

The important magnetic parameters like saturation magnetization, remanent magnetization and coercivity were calculated from the hysteresis loop and are tabulated in Table 4.3 for all the $LaFeO_3$ nanostructures. Also the ratio of saturation and remanent magnetizations Mr/Ms was calculated for each morphology to determine the type of magnetic domain present in the prepared nanostructures.

Table 4.3. Magnetic Parameters of the different Morphologies of $LaFeO_3$ Nanostructures

Nanostructures	*Saturation Magnetization (M_s)(emu/g)*	*Remanent Magnetization (M_r) (emu/g)*	*Coercivity (Oe)*	*Mr/Ms*
Nanocubes	6.47	0.640	226	0.09891
Flowers	1.924	0.353	431	0.1837
Dendrites	4.03	0.845	205	0.2096
Nanorods	3.86	0.825	198	0.2137
Microspheres	2.60	0.062	85	0.0231
Nanospheres	1.02	0.006	14	0.0062

On the basis of the proposed criteria [18], the nature of domain in a ferromagnetic material of a particular size the can be determined, by calculating the Mr/Ms value from the hysteresis loop. According to that hypothesis, single domain (SD) particle have a Mr/Ms value of large than 0.5, while the pseudo-single-domain (PSD) particle have a value between 0.1 to 0.5 and multidomain (MD) particle have a value lower than 0.1.Figure 4.14 a-d shows the M-H loop recorded from the different nanostructures like nanocubes, flowers, dendrites and nanorods respectively. Among all the prepared nanostructures in the present study, the nanocubes were found to exhibit a high magnetization and remanenance. The Mr/Ms values of 0.0.09891 were calculated from the M-H loop of the nanocubes suggests the presence of multidomain (MD) behaviour.

Similarly, the Mr/Ms value calculated for the M-H loop of the other nanostructures like flowers, dendrites and nanorods shows the values of 0.1837, 0.2096 and 0.2137, respectively again suggesting the presence of all nanostructures pseudo-single domains (PSD).The variation in Mr/Ms value can be related to the average particle size and morphology which related to the use of surfactant in the reaction, as already discussed in chapter 3. The nanocubes prepared without surfactant, another morphologies like flowers, dendrites and nanorods prepared with the surfactants of PVP, CTAB and citric acid, which controls the initial grain growth, shape and smaller particles size. The reduction of particle size with different surfactant in the synthesis of flowers, dendrites and nanorods, might have led to the formation of PSD particles. The above results suggest that the nature of surfactant play major role in determining the magnetic properties of the samples.

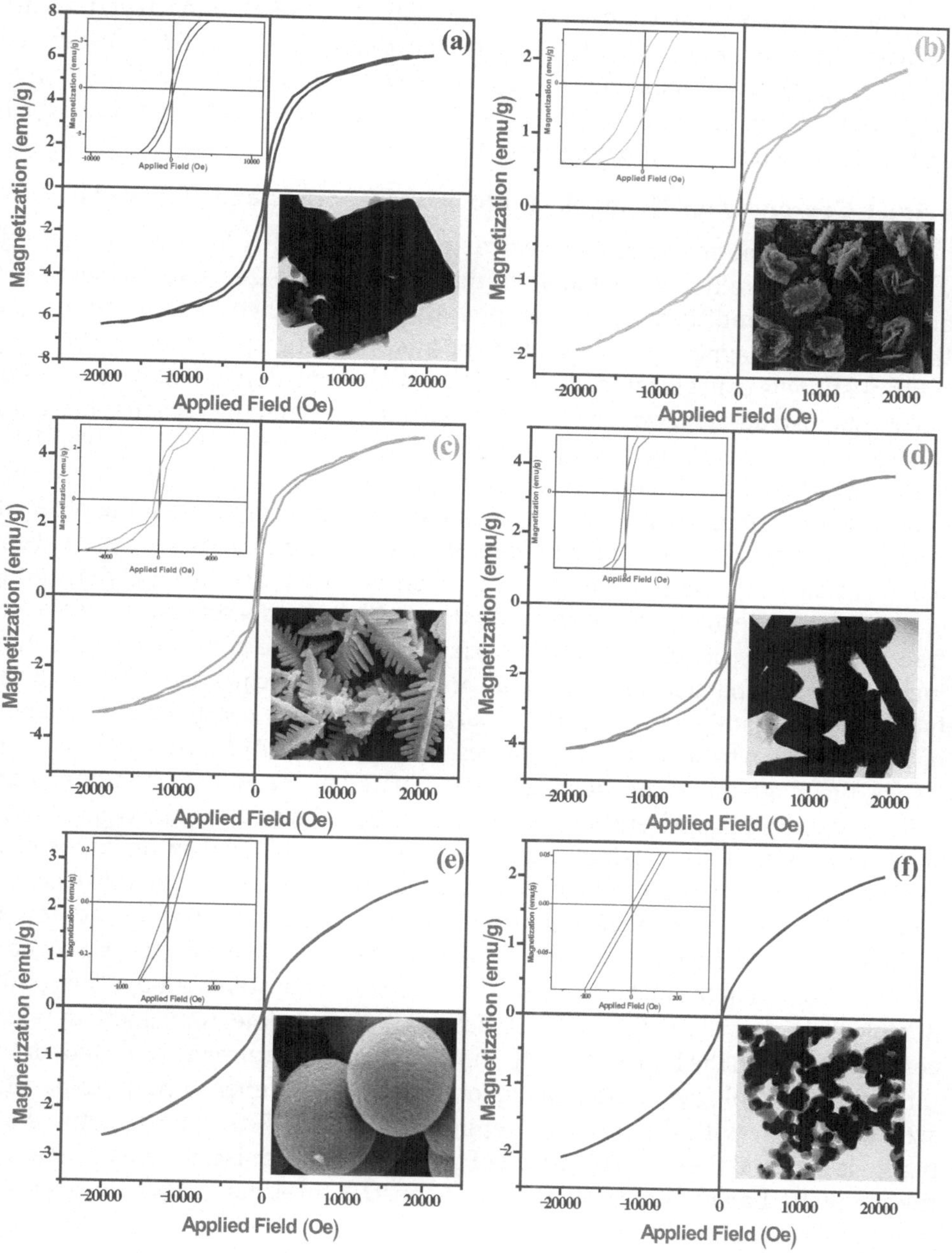

Figure 4.14. Room Temperature M-H loop of the different $LaFeO_3$ Nanostructures and Inset (Left) Shows the Magnified View of the M-H Loop at Low Applied Magnetic Field (Right) the Corresponding Nanostructures.

The M-H loop of the sphere like morphologies of $LaFeO_3$ nanostructures are shown in Figure 4.14 (e and f). The Mr/Ms value of 0.0231 and 0.0062 are estimated for microspheres and nanospheres respectively, and it suggest the formation of multidomain. The presence of MD is also reflected in the coercivity value, which will be discussed in the following section.

4.4.4.1 Coercivity for Nanocubes, Flowers, Dendrites and Nanorods

All the magnetic parameters extracted from the M-H loop, the coercivity values is in generally high sensitive to the variations in size, shape, aspect ratio and synthesis conditions of the sample. Also, the coercivity values determine the use of the magnetic materials for specific magnetic applications. Since the present study is concentrated about the size and shape dependent properties of $LaFeO_3$ nanostructures, we look an initiative to understand the reason underlying the variation of coercivity among the different morphologies of $LaFeO_3$. The M–H curves measured for $LaFeO_3$ nanocubes are shown in Figure 4.14. The results shows that the magnetization of $LaFeO_3$ nanospheres is larger than the other all $LaFeO_3$ morphologies. The magnetic moments of the antiferromagnetic $LaFeO_3$ nanoparticles originate from the spin uncompensation at the surface and the canting of the Fe-Fe antiferromagnetic order. As the particle size decreases, an increasing fraction of atoms lie at or near the surface and then the spin uncompensation increase. So, the particle moment of an antiferromagnetic particle is expected to increase significantly with decreasing particle size. The Fe cation via the oxygen p orbitals is known to be the mechanism responsible for the AFM ordering in $LaFeO_3$. Hence, the magnetic properties are intimately tied to the Fe-O-Fe bonding of the FeO_6 octahedra [19], where the Fe cations are coordinated to six oxygen atoms. Distortions of the FeO_6 octahedra may result from the epitaxial strain imposed by the substrate, thus affect the magnetic properties. Here, the magnetization of $LaFeO_3$ nanocubes are larger than the other $LaFeO_3$ morphologies, which can be attributed to the size of the $LaFeO_3$ nanocubes is large than the other nanostructures. Secondly, it can be found that the all $LaFeO_3$ nanostructures show hysteresis loop. The results demonstrate that the $LaFeO_3$ nanostructures show weak ferromagnetism which corresponds with the bulk form [20]. Similar results were already reported for the different morphologies antiferromagnetic $LaFeO_3$ nanotubes with finite particle size. Compared with the previous reports [20], the $LaFeO_3$ nanostructures are bigger in size and they are composed of grains with finite size. So, we deduce that the magnetic behaviours of the $LaFeO_3$ nanostructures may originate from the shape and finite size effect.

Next, it can be seen from the coercivity value shows significant variation from one morphology to another. Interestingly, the coercivity value calculated from Figure 4.15 a cis 226 Oe from the hysteresis loop of $LaFeO_3$ nanocubes. It was also observed that, there was a variation in coercivity value between the flowers (431 Oe), dendrites (205 Oe) and nanorods (198 Oe) as shown in Figure 4.15 b-d. Generally, the presence of surfactant is claimed to be one of the factors affecting the magnetic properties of the sample. The influence of the surfactant can be of two

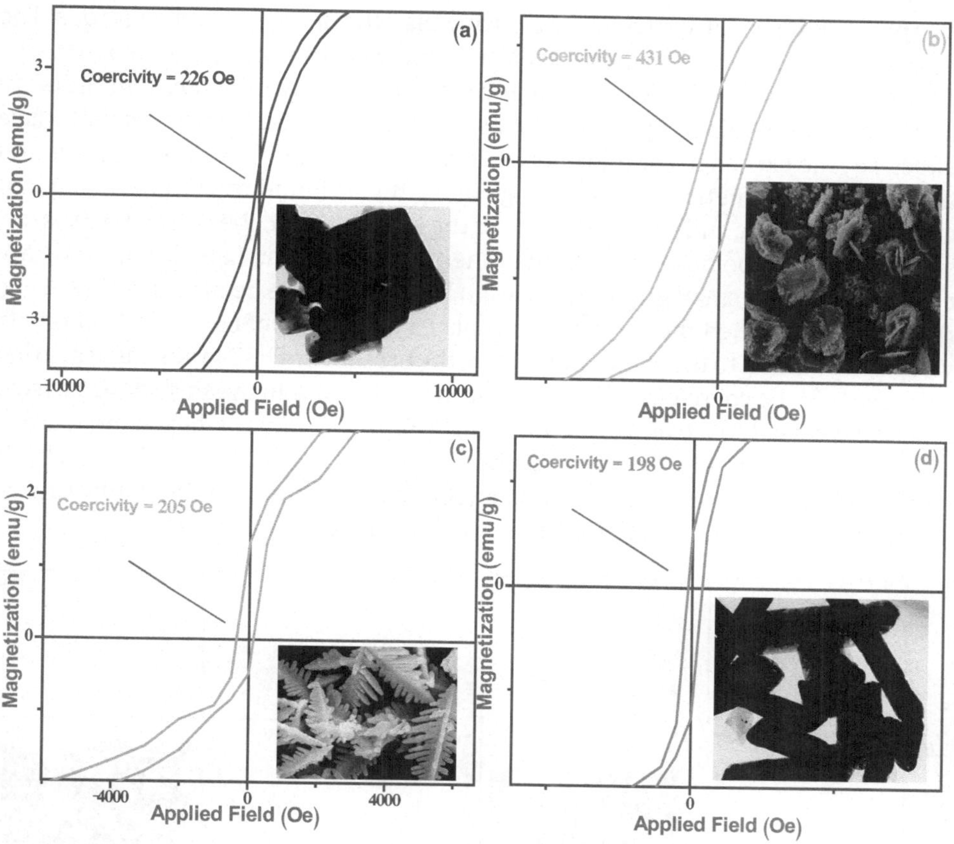

Figure 4.15. Magnified View of the M-H Loop at Low Applied Field and Inset Shows the Corresponding $LaFeO_3$ Nanostructures (a) Nanocubes, (b) Flowers, (c) Dendrites and (d) Nanorods.

types. The more presence of surfactant can lead to the formation a non-magnetic dead layer on the surface of the sample, which decrease the overall magnetization in the sample. In certain cases, the surfactant acts as structure directing agent and control the growth of different morphologies and finally get removed completely during the reaction, at elevated temperatures and subsequent washing of the samples. In such cases, the surfactants do not directly influence the magnetic properties, rather, the unique morphologies formed in the surfactant controlled growth that exhibits size and shape dependent magnetic properties.

In the present study, the presence of the surfactant was not observed in any of the analysis. Hence, shape anisotropy observed might be the sole factor responsible for the variation in magnetic parameters. This shape dependence of magnetic behaviour is more pronounced when the particle deviates from the usual spherical shape to anisotropic morphologies. It has been well established that, the coercivity increase with increase in shape anisotropy. So, in our present case, flower like morphology, has relatively more anisotropy than the dendrites and nanorods. The flower like structures which poses even more increased anisotropy compared to the dendrites and nanorods had still higher coercivity value. No obvious loop shifts are found in other obtained hysteresis loops. Loop shift originates from the interface effect, *i.e.*, the exchange coupling between the spins of the disordered surface and the "antiferromagnetic" core (exchange bias). It was found to be very dependent on the particle size and shape. The size of the $LaFeO_3$ nanocubes is relatively large, so loop shift can be observed. On the other way, due to the $LaFeO_3$ nanorods are composed of very small grains, there is no loop shifts have been found. The above results we confirmed the morphologies play a major role of controlled coercivity values. In addition to the shape anisotropy, the presence of PSD grain might also led to increase in the coercivity. The reduction in particle size in the surfactant assisted reaction can lead to a transformation from MD to PSD range, which contribute to the increased coercivity. Thus the relatively high coercivity of flower like structures compared to dendrites and nanorods might be due to the combined influence of shape anisotropic and smaller particles in PSD range.

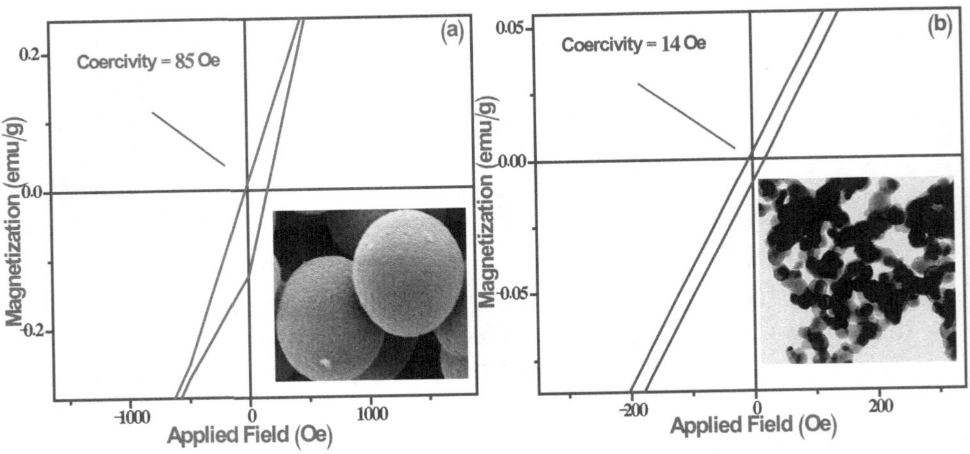

Figure 4.16 Magnified View of the M-H Loop at Low Applied Field and Inset Shows the Corresponding $LaFeO_3$ Nanostructures (a) Micropheresm, (b) Nanospheres.

4.4.4.2 Coercivity in Microspheres and Nanospheres

The M-H loop of another two morphologies, namely microspheres and nanospheres shows less coercivity values of 85 and 14 Oe as shown in Figure 4.16 a and b. We speculate that the surface spins may array at the surface of all $LaFeO_3$ nanostructures at room temperature. The nanocubes, flowers, dendrites and nanorods like morphologies are helpful to the surface spins to form net moment while microspheres and nanosphere morphology is not (shape anisotropy). So, obvious surface effect (surface anisotropy) can be observed for all other four $LaFeO_3$ nanostructures. However, the M(T) and M(H) curves of the prepared $LaFeO_3$

micro and nanospheres exhibits weak ferromagnetism. It may be caused by the antiferromagnetic order with canted spins [19]. In addition, at room temperature couples of Fe^{3+}- Fe^{2+} may be appeared in $LaFeO_3$ due to losing of oxygen.

The difference between magnetic moment of Fe^{3+} ions (5 μB) and Fe^{2+} (4 μB) has contributed to magnetic behaviours of the samples and they became an electrical conducting material as semiconductor. [20] If the prepared micro and nanospheres has some of particles with multiple domain sizes, thereby Hc, Mr, and Mr/Ms will differ from zero. The weak ferromagnetic behavior is more clear with the small particle size gives lower Mr/Ms. That is the reason we suggested the ratio Mr/Ms could be used as a functionally parameter for evaluating the homogeneity on dimension of nanoparticles and the limit of multidomain size of the magnetic materials.

The large coercivity value for the nanocubes, flowers, dendrites and nanorods might be due to the combined effect of

- ☆ Shape anisotropic induced due to the deviation from micro and nanospheres
- ☆ Small particles in PSD range as reflected from the Mr/Ms value calculated from the M-H loop and
- ☆ Surfactant used in the prepared of these nanostructures, which might have resulted in OH groups and associated surface spin pinning.

4.5 Electrical Properties of different $LaFeO_3$ Nanostructures

4.5.1 Introduction

Complex perovskite oxides have been of great importance since a number of materials exhibiting a wide range of unusual properties. These include magnetic materials, superconductors, laser host materials and dielectric materials. The ability of the simple perovskite structure to give rise to intergrowth structures, oxygen deficient structures and ordered perovskite structures that exhibit both electronic and ionic conductivity have numerous technological applications, such as electrodes for solid oxide fuel cells, gas semi-permeation membranes, water electrolysis, and electrocatalytic reactions. [21] It has been reported that some perovskites with formula ABO_3, where A is an alkali rare earth like La, Ca, Sr, Ba, *etc.*, B is an element with valence +3 or +2, such as Fe and Ni, exhibit quite high proton and oxygen ion conductivity [22]. It is expected that mixed ionic and electronic conductors may be found in these compounds if the B-sites are partially substituted by the appropriate first row transition elements. It has been reported that Fe-containing perovskite exhibits quite high O^{2-} ion and electronic conductivity [23].

The impedance analysis has been widely used to study the dielectric behaviour of the crystalline and polycrystalline ceramic materials. In general, the dielectric properties of ferroelectromagnetic materials arise due to intra-grain, inter-grain and other electrode effects. The motion of charges could take place by charge displacement, dipole reorientation, space charge formation *etc.* [24]. In order to understand the electrical properties of a given sample grain, grain boundary and electrode contributions must be separated out. Complex impedance analysis has emerged as a very powerful tool for separating out these contributions. Similar analyses on different types of perovskite oxide have been reported earlier [24].

Several reports are available in the literature [23, 25] on the electrical conductivity and dielectric properties of perovskite oxide. However, to our knowledge reports are not available, for the nanostructured $LaFeO_3$ perovskite oxide in general and in particular morphology dependent electrical properties of $LaFeO_3$. This has motivated us to carry out electrical and dielectric properties of $LaFeO_3$ nanostructure. The present work deals with the impedance measurements and also dielectric properties of nanostructure $LaFeO_3$ with various morphology.

4.5.2 Experimental Procedure

The impedance measurements were performed for all the samples. Different morphologies of $LaFeO_3$ nanostructured powder was made into pellets of 8 mm diameter and ~ 2 mm thickness for the impedance measurements. AC impedance spectroscopic measurements have been carried out using a Hewlett–Packard HP 4284A model Precision LCR Meter over the frequency range from 1 MHz to 20 Hz in the temperature domain 300 – 340 K. The electrical conductivity value was estimated using the relation $\sigma = t/(R_bA)$, where t is the thickness of the specimen, A the area covered by the silver electrodes in contact with the specimen and Rb is the bulk resistance of the material derived from the intercept of the complex impedance plot *i.e.*, Nyquist plot on the real axis. The equivalent circuit software developed by Boukamp et. al. [26] has been effectively used to extract the accurate value of grain boundary from the intercept on the real axis at the high frequency end of the relevant Nyquist plot.

4.5.3 Results and Discussion

4.5.3.1 Conductivity Analysis

Impedance measurements were carried out for all $LaFeO_3$ nanostructures such as nanocubes, flowers, dendrites, nanorods, microspheres and nanospheres at selected temperatures in the frequency range 20 Hz–1 MHz. Figure 4.17 (a) shows the impedance plot for $LaFeO_3$ dendrite sample at RT. This consists of two semicircular arcs. The arc on the low-frequency side is due to the grain boundary conduction and that on the high-frequency side is due to the grain conduction.

The parameters R_g, C_g and $(\omega p)_g$ correspond to the resistance, capacitance and the relaxation frequency $(=1/\tau)$ of the grain and R_{gb}, C_{gb} and $(\omega p)_{gb}$ are the corresponding quantities for the grain boundary. These parameters were obtained for each temperature by analysing the data using the non-linear least square (NLLS) fitting. The resistance values *Rg* and *Rgb* were used to calculate the dc conductivity of the grain and the grain boundary respectively. For the $LaFeO_3$ samples semicircles corresponding to the conduction due to grain and grain boundary were obtained suggesting a predominant contribution from the grain boundary is not well resolved in the present study. Hence we presume that the conductivity of all samples is mainly due to the grain boundary contribution. Figure 4.17 (b) reveal the measured complex impedance spectra for $LaFeO_3$ dendrite at five different temperatures. Similar measurements have also been carried out for other samples. The relaxation frequency (ω_p) for the grain boundary shifts to lower values with the decrease in crystallite size of the various morphology as shown in Table 4.4.

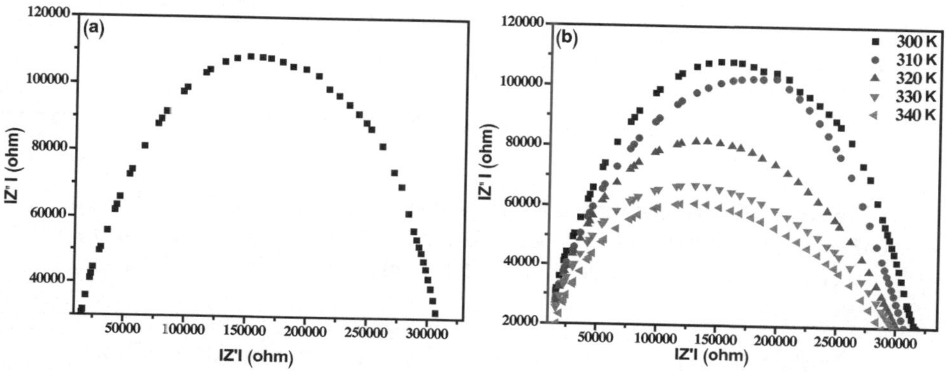

Figure 4.17. Complex Impedance Spectra for $LaFeO_3$ Dendrite Samples (a) Room Temperature, (b) different Temperature.

Table 4.4. Relaxation Frequencies of Various $LaFeO_3$ Nanostructures at different Temperatures

Temperature (K)	*Nano-cubes ω_p (Hz)*	*Flowers ω_p (Hz)*	*Dendrites ω_p (Hz)*	*Nanorods ω_p (Hz)*	*Micro-spheres ω_p (kHz)*	*Nano-spheres ω_p (kHz)*
300	1740	1810	6808	7012	278	286
310	1207	1342	5177	5602	214	232
320	869	989	4045	4246	184	194
330	693	812	3286	3541	161	184
340	496	605	2468	2613	139	152

From Figure 4.18a, it can be observed that the conductivity and its temperature dependence for the grain are quite different from that of the grain boundary for all $LaFeO_3$ nanostructure samples. The activation energies for the grain conduction are found to be in the range of 0.17 and 0.30 eV for different $LaFeO_3$ nanostructures, which are smaller to those of the grain boundaries and their values are given in Table 4.5. The values of activation energies have the characteristic of electron hopping between Fe^{2+} and Fe^{3+} ions in *B* sites. For $LaFeO_3$ microspheres and nanosphere, there is a change in the slope of the Arrhenius plot in Figure 4.18a at about 310 and 320 K, respectively.

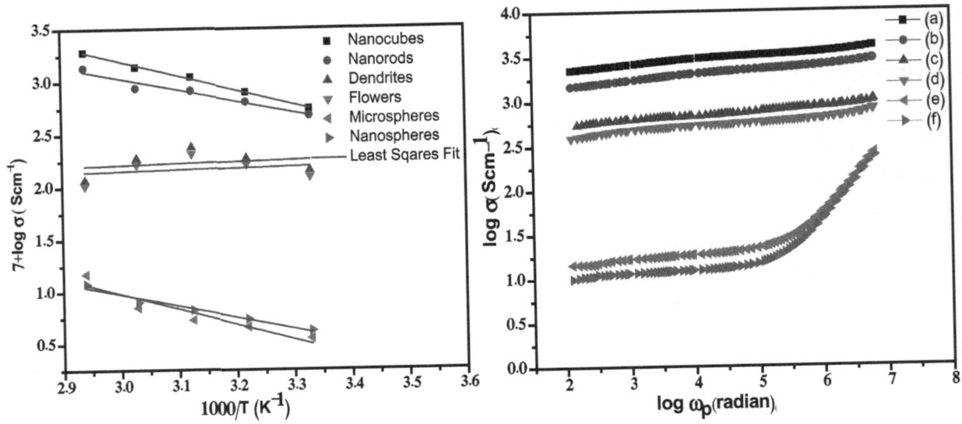

Figure 4.18. (a) Arrhenius Plots of the Conductivity of the different $LaFeO_3$ Nanostructures. The solid lines are the best fit to equation and (b) Conductance Spectrum of different $LaFeO_3$ Nanostructure.

Table 4.5. Conductivity Values Evaluated from Impedance Plots for the different $LaFeO_3$ Nanostructures at 300 K and their Corresponding Activation Energies

Nanostructure	*Conductivity σ (Scm⁻¹)*	*Activation Energy (eV)*
Nanocubes	3.58×10^{-7}	0.171
Flowers	3.42×10^{-7}	0.215
Dendrites	5.54×10^{-5}	0.222
Nanorods	4.75×10^{-5}	0.235
Microspheres	1.431×10^{-5}	0.287
Nanospheres	1.329×10^{-5}	0.303

Below these temperatures, the conductivity can be explained by the same mechanism of e^- hopping between Fe^{2+} and Fe^{3+} ions as in the case of the other four $LaFeO_3$ nanostructures. The higher values of activation energies, obtained for the high temperature conductivity, suggest that the conduction mechanism in this temperature range should be different from that in the low temperature region. It was already reported that oxygen ions escape from the perovskite structure and thereby create anion vacancies [27]. Therefore, the conduction due to anion vacancies may be responsible for the conductivity behavior with higher activation energy (~ 0.30 eV) in the high temperature region for micro and nanosphere samples, which also have small grain size. The value of this activation energy has the characteristic of diffusion of oxygen vacancies. The other $LaFeO_3$ nanostructure the grain size range 90-70 nm (as discussed in Chapter 3) the number of oxygen vacancies is expected to be negligible so that conduction is mostly due to *n*-type carriers only. Single activation energy is obtained for all samples as the electron mechanism shows semiconductor regions with two semiconducting regions with two different activation energies as in Figure 4.18b. In Table 4.5 it can be observed that these activation energies (given in parentheses) agree with those obtained from the complex impedance measurements.

Figure 4.18b shows a plot of log ω_p vs log (σ) for various morphology of $LaFeO_3$ nanostructures. The values of ω and σ are obtained from the best fits using NLLS fitting. The solid lines are the least squares straight line fits, all of which give a slope of almost unity implying that the dc and ac conductions are correlated to each other in $LaFeO_3$ nanostructures for various morphology and that they are governed by the same mechanism.

4.5.4 Dielectric Behaviour

4.5.4.1 Frequency Dependence

$LaFeO_3$, the rotation of Fe^{2+} - Fe^{3+} dipoles may be visualized as the exchange of electrons between the ions so that the dipoles align themselves with the alternating field. The existence of inertia to the charge movement would cause a relaxation to the polarization. The effects of frequency (ω) on the dielectric constant ε' and dielectric loss factor tan δ (at 300 K) for samples with various morphology of $LaFeO_3$ are illustrated in Figure 4.19 (a) and (b) respectively. In our measurements ε' for $LaFeO_3$ nanocube, flowers, dendrite and nanorods is found to be one orders of magnitude smaller than the $LaFeO_3$ micro and nanospheres. Similarly the dielectric loss is also found to be one order of magnitude smaller for the micro and nanospheres. The existence of Fe^{3+} ions and the minority Fe $^{2+}$ ions have rendered $LaFeO_3$ dipolar. Rotational displacements of the dipoles result in orientational polarization

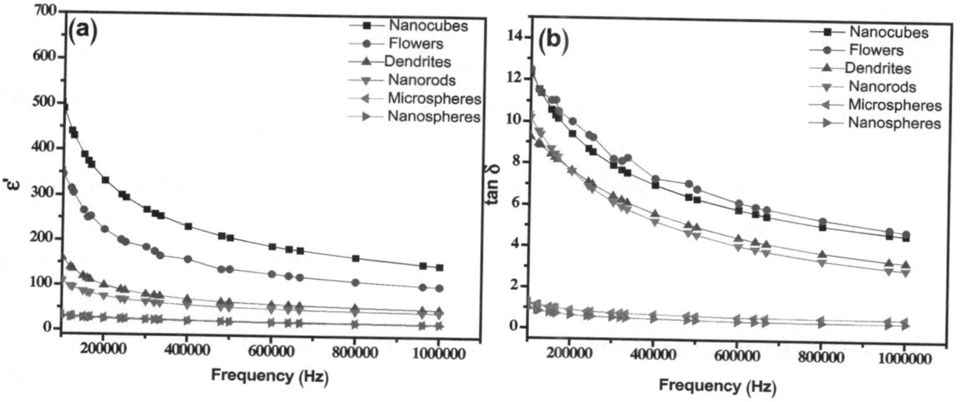

Figure 4.19. (a) Real Part ε' of the Dielectric Constant at 300 K, (b) Dielectric Loss Factor tan δ at 300 K as a Function Frequency for the different $LaFeO_3$ Nanostructures.

The dielectric constant ε' shows an anomalous behaviour with frequency for the $LaFeO_3$ nanostructures. It initially increases with frequency, reaches a maximum and then decreases with further increase in frequency. The value of ε' decreases and the peak position shifts to a lower frequency on grain size reduction. In the normal behaviour, the polarization (or dielectric constant) will monotonically decrease with frequency as in the case of nanocube and flowers. This happens as the electronic exchange between Fe^{2+} depu134 Fe^{3+} cannot follow the alternating field beyond a certain frequency. In $LaFeO_3$ the presence of La^{3+} and La^{2+} ions gives rise to p-type carriers. The local displacements of p carriers in the direction of the external electric field also contribute to the net polarization in addition to the n-type carriers. However, the contribution of the p-type carriers should be smaller than from the electronic exchange between Fe^{2+} depu134Fe^{3+} and also it has an opposite sign. Since the mobility of the p-type carriers is smaller than the n-type carriers, the contribution to polarization from the former will decrease more rapidly and even at lower frequencies than from the latter. As a result the net contribution will increase initially and then decrease with frequency as observed in the present samples. The shift in the position of the maximum towards lower frequencies when the grain size is reduced can be explained by proposing a decrease in the number of p-type carriers in the B sites as has been reported [27]. The real part of the dielectric constant decreases with frequency and also the loss is lower for smaller grain size as the resistivity increases with grain size reduction.

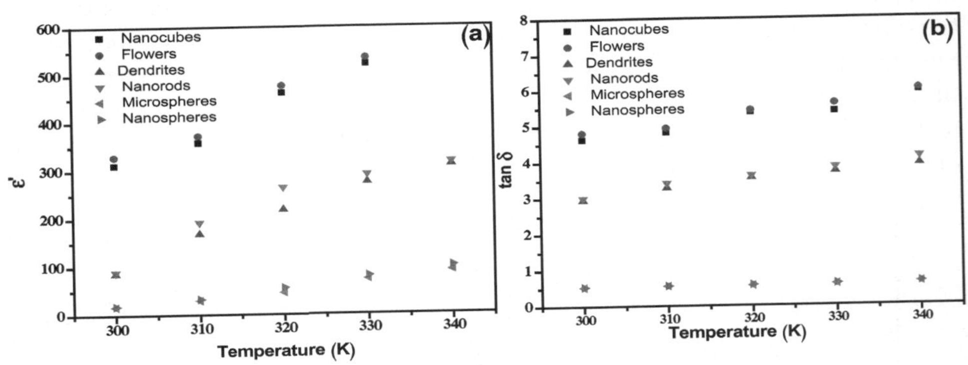

Figure 4.20. (a) Real Part ε' of the Dielectric Constant at 1000 Hz, (b) Dielectric Loss Factor tan δ at 1000 Hz as a Function Temperature for the different $LaFeO_3$ Nanostructures.

4.5.4.1 Temperature Dependence

The anomalous temperature dependence of ε' for different $LaFeO_3$ nanostructures, as seen from Figure 4.20 (a), could be explained as in the case of its frequency dependence. The contributions of the two types of carrier to polarization depend on temperature. Since the temperature can influence the electronic exchange of Fe^{2+} depu134 Fe^{3+} more than it could the displacements of p-type carriers. The ε' will increase with temperature at the beginning, but above a certain temperature the contribution from the p-type carriers also becomes significant and thus ε' will begin to decrease since the two contributions oppose each other. The dielectric losses, as seen from Figure 4.20 (b), increase as a result of the decreasing resistivity of the samples with temperature. Also, the loss decreases with grain size reduction.

4.5 Conclusion

In summary, the important physical properties of $LaFeO_3$ nanostructures were quantitatively observed. The specific surface area, optical properties, magnetic and electrical properties were highly dependent on the size and shape of the nanostructures. This assures the surface properties can be easily tuned for a specific application, by judicious control on the surfactant and growth condition. The optical properties of different $LaFeO_3$ nanostructures were studied. It was found that the absorption cut-off wavelength varied depending on the morphologies. The absorption edge at 580 - 600 nm observed using UV-Vis DRS emphasis the fact that all $LaFeO_3$ nanostructures could serve as a potential visible-light-driven photocatalytic material. Similarly, the detailed analysis on the magnetic properties of the different nanostructures of $LaFeO_3$ provides sufficient information on the shape dependent magnetic properties. Finally the electrical behaviour of different $LaFeO_3$ nanostructures was studied. It was found that impedance and electrical conductivity also shows disparity depending on the morphology. The next step of the work is to utilize these novel shape dependent physical properties to explore the multifunctionality of theses nanostructures in variety of applications. The forthcoming chapter V of this thesis deals with the study of the photocatalytic and biosensing performance of the different $LaFeO_3$ nanostructures.

References

[1] Y. Masuda and K. Kato, *Chem. Mater.,* **20** (2008) 1057-1063.

[2] J. Goldberger, R. Fan and P. D. Yang, Acc. *Chem. Res.,* **39** (2006) 239-248.

[3] J. H. Su, L. Q. Jing, K. Shi, C. H Yao H. G. Fu, *J. Nanopart. Res.,* **12** (2010) 967-974.

[4] W. Zhou, K. Tang, S. Zeng, Y. Qi, *Nanotechnology,* **19** (2008) 065602.

[5] K. A. Gregg, S. C. Perera, G. Lawes *et al., Chem. Mater.,* **18** (2006) 36.

[6] K.S. W Sing, D. H Everett, L. Moscou *et al., Pure and appl. Chem,* **67** (1984) 603.

[7] L. Peisheng, *Chinese J. Mater. Res.* **23** (2009) 415-421.

[8] A. Niazi, P. Poddar, A. K. Rastogi, *Current Science.* **79** (2000) 99-104.

[9] H.G. Yang, H.C. Zeng, *J. Phys. Chem. B* **108** (2004) 3492–3495.

[10] J. Ding, X. Lu, H. Shu, J. Xie, H. Zhang., *Mater. Sci. Eng. B* **171** (2010) 31-34.

[11] J. Chandradass, K. H. Kim, *et al.,* Mater. *Chem. and Phy.* **122** (2010) 329 -322.

[12] M. Rajendran and A. K. Bhattacharya, *J. Eur. Ceram. Soc.* **26** (2006) 3675

[13] A. A. Cristobal, P. M. Botta, P. G. Bercoff, *Mater. Res. Bull.* **44** (2009) 1036.

[14] R. H. Kodama, S. A. Makhlouf, and A. E. Berkowitz, *Phys. Rev. Lett.* **79** (1997)1393.

[15] E. Winkler, R. D. Zysler, M. V. Mansilla, *Phys. Rev. B* **72** (2005) 132409.

[16] H. W. F. Sung, C. Rudowicz, *J. Magn. Magn. Mater.*, **312** (2007) 250-256.

[17] S. Zeng, K. Tang, T. Li, *J. Coll. Inter. Sci*, **312** (2007) 513-519.

[18] D. Wang and M. Gong, *J. Appl. Phy.* **109** (2011) 114304.

[19] A. Hossein, S. Hadi, A. Poddar *et al.*, *J. Phys. D: Appl. Phys.* **43** (2010) 245002 (5pp)

[20] E. Folven, T. Tybell, A. Scholl, A. Young, *et al.*, *Nano Lett.* **10** (2010) 4578–4583.

[21] P. Jha, S. Bobev, G. N. Subbanna and A. K Ganguli., *Chem. Mater.* **15** (2003) 2229.

[22] S.S. Chougule and B.K. Chougule, *Smart Mater. Struct.* **16** (2007) 493.

[23] R.S. Devan and B.K. Chougule, *J. Appl. Phys.*, **101** (2007) 014109.

[24] L. Sonali, S.S Darshane, Survanshi, I.S.Mulla, *Ceramic International*, **35** (2009) 1795.

[25] N. Ponpian, P. Balaya and Narayansamy A, J. *Phys.,Conden. Mater.* **14** (2002) 3221.

[26] Y. Zhi and A. Chen, *J. Appl. Phys.* **91** (2002) 794.

[27] G.F. Goya, H.R. Rechenberg, *J. Magn. Magm. Matter.* **203** (1999) 141.

Chapter 5

Morphology Dependent $LaFeO_3$ Nanostructures for Photocatalytic and Biosensing Application

PART- I

5.1 Morphology Dependent Photodegradation Study of $LaFeO_3$ Nanostructures for Rhodamine B under Visible Light

5.1.1 Introduction

In recent years, nanomaterials based photocatalytic degradation has been known to be a promising method to solve the environmental and energy crisis. Photocatalytic degradation of organic pollutant, such as Rhodamine B (RhB) [1, 2], is difficult due to the stable dyestuff. It has been proved to be a viable environmentally benign technology to the decolouration and degradation of dye effluents. Within various categories of photocatalysts, perovskite-type oxides have been found to exhibit high photocatalytic activity to the degradation of organic compounds under visible light irradiation and works in various environments steadily [3]. Therefore, there is a requirement for efficient and environmental friendly methods for decomposition of organic pollutants. With the advent of nanotechnology, environmental remediation using photocatalysis has taken a new dimension. Nanoparticles exhibit high chemical reactivity which is not observed for large particles because of their unusual crystal shape and lattice order. The unique functionality of nanomaterials such as high surface area, easy dispersibility and well defined morphologies have paved way for the proliferation of nanotechnology for environmental remediation.

In the last two chapters, the discussion has been made for the preparation and characterization of different morphologies of $LaFeO_3$ nanostructures and is deliberatively concluded that the morphology and surface properties depends on the surfactant and growth parameters. Based on the obtained results, an attempt has been made to use these $LaFeO_3$ nanostructures with improved surface area for photocatalyst for the degradation of organic effluents. The enhancement in the photodegradation activity with different morphologies with unique surface properties has been quantitatively demonstrated by conducting experiments to degrade a model organic pollutant RhB under visible light irradiation.

5.1.2 Fundamentals of Photocatalysis

Photocatalytic process, as an environmental application is a relatively novel subject with tremendous potential in the near future. Heterogeneous photocatalysis can be defined as the acceleration of photoreaction in the presence of a material catalyst [4]. The term "photocatalysis" is still under debate, since it leads to an incorrect perception that light is acting as a catalyst in the reaction. But instead, light acts as a reactant which is consumed in the chemical process. The actual photocatalyst is a chemical species which may be a metal oxide, perovskite oxide or a mixed oxide. In reality, the term "photocatalysis" does not imply catalysis by light, but rather the 'acceleration of a photo induced or photoactivated reaction in the presence of a material catalyst.

5.1.3 Steps Involved in the Heterogeneous Catalysis Process

Perovskite semiconductor mediated heterogeneous catalytic system is associated with five discrete processes [5]

- ☆ Adsorption of the reactants
- ☆ Reaction in the adsorbed phase
- ☆ Desorption of products
- ☆ Removal of products from the interface region
- ☆ Transfer of liquid or gaseous phase reactants to the catalytic surface

The photocatalytic reaction occurs in the second step of the process *i.e.* the adsorbed phase. The heterogeneous photocatalytic reaction in the adsorbed phase requires, an emitted photon, a catalyst surface (usually a semiconductor material) and a strong oxidation agent (in most case oxygen). In a heterogeneous photocatalysis system, photo induced molecular transformation or reaction takes place at the surface of the catalyst. The photocatalysis process can be generally divided into two classes, depending on where the initial excitation occurs. When the initial photo excitation occurs in the adsorbate molecules which then interact with the ground state catalyst, is referred to as a catalyzed photoreaction. When the initial photo excitation takes place in the catalyst which transfers an electron or energy to a ground state adsorbate molecule, is referred as sensitized photoreaction.

5.1.4 Perovskite Oxide Photocatalytic Materials

Photocatalyst in other words, the catalyst used in the photo assisted reaction, is a solid that can promote reactions in the presence of light, without being consumed in the overall reaction. The catalysts can either provide energy levels to mediate electron transfer between adsorbate molecules (by temporarily accommodating an electron) or behave as both an electron donor (the photogenerated electron in the conduction band) and an electron acceptor (the photogenerated hole in the valence band).

An ideal photocatalyst should be

- ☆ Photoactive
- ☆ Biologically and chemically inert
- ☆ Able to utilize visible light for photoexcitation
- ☆ Inexpensive, Non-toxic

Invariably perovskite oxides have been found to be most suitable photocatalysts because of their photocorrosin resistance and their wide band gap energies. A list of different metal and perovskite semiconductors suitable for photocatalysis and their band gap values is listed in Table 5.1

Table 5.1. Different Photoactive Metal and Perovskite Semiconductors with their Corresponding Bandgap Energy

Semiconductor	*Band Gap (eV)*	*Semiconductor*	*Band Gap (eV)*
TiO_2 (rutile)	3.0	$BiFeO_3$	2.6
SnO_2	3.5	WO_3	2.7
ZnS	3.7	$\alpha - Fe_2O_3$	2.3
$LaFeO_3$	2.1		

In addition to the above-mentioned properties for a semiconductor to be photo chemically active, the redox potential of the photogenerated valence band hole must be sufficiently positive to generate OH radicals, which can subsequently oxidize the pollutant. Also, the redox potential of the photogenerated conductance band electron must be sufficiently negative to reduce the adsorbed O_2 to superoxide. Many factors influence the activity of a photocatalysts in the photodegradation reaction. They are shape, size and surface properties and the preparation spectral activation [6]. Among these parameters, the effect of shape, size and surface properties at the nanoscale have a profound influence, during catalysts. One such example is the suppression of electron-hole recombination, by improving the surface properties at this nanoscale.

5.1.5 Reaction Mechanism of Photocatalytic Reactive RhB

Basically, photodegradation occurs by the generation of electron/hole pairs in the perovskite oxides by absorption of light with energy equal to or higher than its band gap. The generated electrons (e^-) and holes (h^+) migrate to the surface of the catalyst and undergo redox reaction with the adsorbed reactants.

Several factors such as morphology, size, surface area and bandgap energy were addressed to understand the variation of photocatalytic activity of different morphologies of $LaFeO_3$ nanostructures. The heterogeneous photocatalytic process is initiated when a photon with energy equal to or greater than the band gap energy (E_g) of the photocatalyst. When it reaches the surface of the photocatalyst results in molecular excitation. E_g is defined as the difference between the filled valence band and the empty conduction band of the photocatalyst, in the order of a few electron volts. The molecular excitation results in the generation of mobile electrons in the higher energy conduction band (E_{cb}) and positive holes in the lower energy valence band (E_{vb}) of the catalyst, according to the following equation:

$$\text{Photocatalyst} \longrightarrow e^- + h^+$$

Because of the excitation, fate of the separated electron hole can follow several pathways.

The photogenerated electron hole pair may recombine and release the absorbed light energy as heat, with no chemical effect or migrate to the surface of the catalyst and initiate the degradation reactions. The reaction mechanism involved during the photocatalytic activities are schematically illustrated in Figure 5.1. During the degradation process, when the catalyst is irradiated with visible light and having energy greater than the threshold, the photogenerated electrons are transferred to the conduction band (CB) from the valence band (VB) leaving the positive holes in the VB. The photoinduced holes in the VB directly oxidizes the pollutants or they are trapped by surface hydroxyl groups at the catalyst surface to yield hydroxyl radicals (OH•) [7].

These effects would improve the oxidising ability of the nanostructures towards the conjugated structure of RhB pollutant. The electrons in the CB involves in the reduction process, which can be attributed to the band gap and electron-hole pair formation in the valence band formed by the hybrid orbitals in the conduction band. The generated conduction band electrons react with dissolved oxygen molecules yielding superoxide anion radicals ($O_2•^-$) which generates the hydroperoxy radicals ($HO_2•$) on protonation. Further, hydroxyl radicals (OH•) are produced which acts as strong oxidising agent to decompose the organic dye into non-toxic products [8]. However, the electron-hole recombination was unavoidable. In the proposed photocatalytic degradation mechanism, higher the specific surface area, more oxygen molecules were absorbed on the catalyst surface producing large number of OH•. On the other hand, the photocatalytic activity was enhanced by the faster transport of the photogenerated carriers. Therefore, the $LaFeO_3$ nanostructures also exhibit different photocatalytic activities.

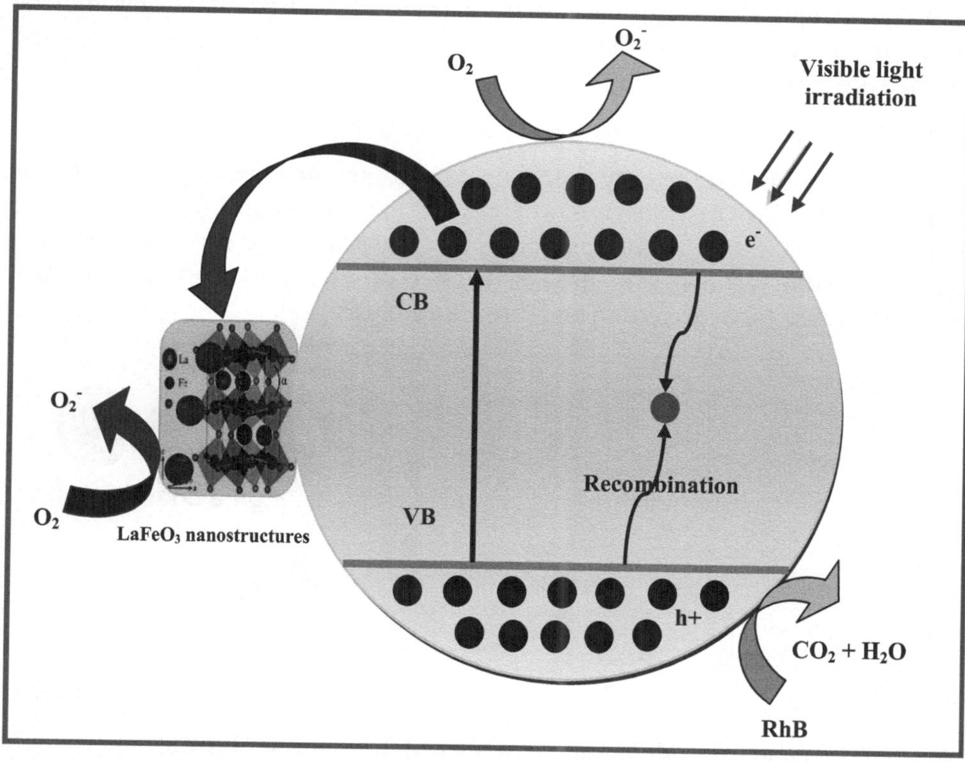

Figure 5.1. Schematic Illustration of the Reaction Mechanism Involved during Photocatalytic Activity.

5.1.6 $LaFeO_3$ as Visible Light Photocatalyst

The semiconducting perovskite $LaFeO_3$ nanostructures act as efficient visible light photocatalyst due to its narrow band gap energy. Generally, the photodegradation is initiated by generation of electron/hole pairs in the semiconductor upon absorption of UV light with energy equal to or higher than its band gap. The generated electrons (e^-) and holes (h^+) migrate to the surface of the catalyst and undergo redox reaction with the adsorbed reactants. The high oxidative potential of the holes at the surface of the catalyst oxidizes the toxic organic compounds at the surface and the surrounding environment of the catalyst to non-toxic substances [9]. Semiconducting materials such as TiO_2, ZnO and ZnS show high photosensitivity. Despite their remarkable features, there are some drawbacks pertaining to faster charge carrier recombination and activation only under ultraviolet irradiation due to their large band gap [10]. So, it is of considerable interest to identify a material of optimal band gap that can respond to a wide range of wavelengths, both in ultraviolet and visible range. The properties and catalytic activities of $LaFeO_3$ strongly depend on the crystallinity, size, surface morphology, optical band gap and preparation methods which dramatically improve its photocatalytic activity. In the following section, we discuss the morphology and surface properties dependent photocatalytic activities.

5.1.7 Measurement of Photocatalytic Activity

The photocatalytic activity was evaluated against the organic dye Rhodamine B (RhB) ($C_{28}H_{31}CIN_2O_3$), which is one of the most important xanthenes dyes used as biological stain, tracing agent, solar collector, paper and laser dye. The molecular structure of RhB is shown in Figure 5.2. It has become a common organic pollutant and resistant to biodegradation and direct photolysis [11].

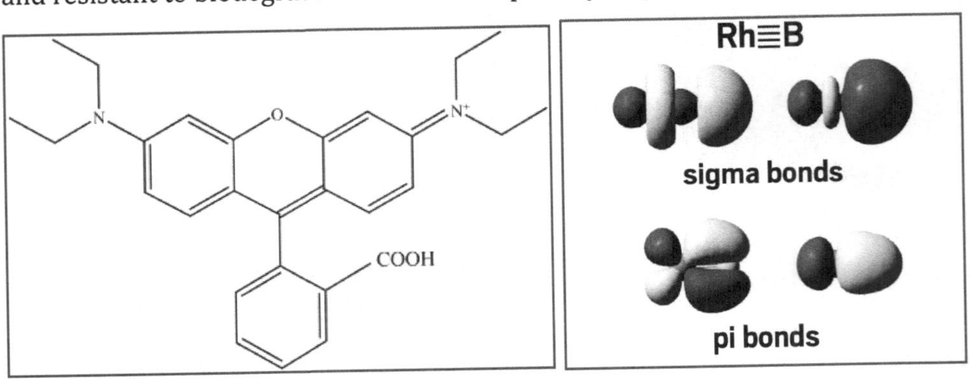

Figure 5.2. Molecular Structure of Rhodamine B (RhB).

The evaluation of photocatalytic activity of the prepared six different morphologies $LaFeO_3$ nanostructures for decolorization of RhB aqueous solution was performed at ambient temperature. 100 mg of the prepared nanostructures was dispersed in 100 mL RhB aqueous solution with a concentration of 5×10^{-5} M. The solution was allowed to reach an adsorption-desorption equilibrium among the photocatalyst, RhB, and water before the visible light irradiation. A 150 W visible light lamp, with the wavelength range > 400 nm was used as light source. A photograph of multi lamp photocatalysts reactor is shown in Figure 5.3. In a typical process, aqueous suspension of RhB (1g/1L) and 100 mg of $LaFeO_3$ were placed in a vessel. Prior to irradiation, the suspensions were magnetically stirred in dark for 30 min to ensure the equilibrium of the working solution. The suspensions were kept under constant air-equilibrated condition before and during the irradiation. The pH of reaction suspension was not adjusted. Analytical samples (3 ml) were drawn from the reaction suspension every 30 min, and removal of $LaFeO_3$ by centrifugation. The changes of absorption at 554 nm were applied to identify the concentration of RhB using UV-spectroscopy (Lambda 35, Perkin-Elmer). Absorption spectrum was obtained by a plot between absorbance versus irradiation time and the efficiency of decolourization was determined in terms of change in intensity of absorption maximum of the dyes.

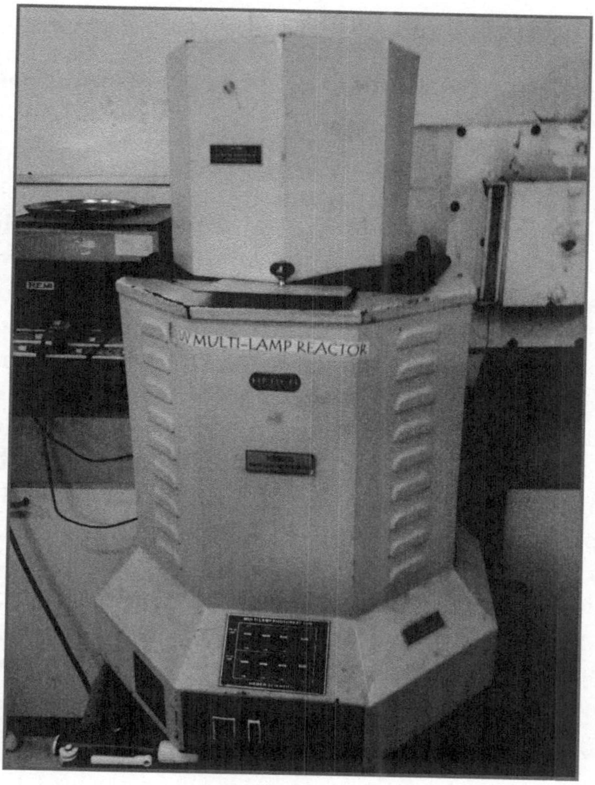

Figure 5.3. Photograph of Multi Lamp Photocatalyst Reactor.

The efficiency of decolourization was calculated using the formula

$$\frac{A_0 - A_t}{A_0} X\ 100$$

Where, A_0 is the absorption maximum at t = 0, A_t is the absorption maximum after complete degradation. In addition to that, the apparent rate constant of photocatalytic degradation of RhB was calculated from the expression

$$\frac{In\ A_0}{A_t} = kt$$

Where, k is the rate constant of the reaction, A_0, A_t are the absorbance maximum at t = 0 and t = min time required to complete the degradation respectively. The apparent rate constant was determined from the slop of the plot between A_0/A_t and irradiation time.

5.1.8 Results and Discussion

5.1.8.1 Photodegradation Studies of different Morphologies of $LaFeO_3$ Nanostructures

In the present work, photocatalytic activities of six different morphologies of $LaFeO_3$ nanostructures such as nanocubes, flowers, dendrites, nanorods, microspheres and nanospheres were performed by degrading the RhB dye and the results are shown in Figure 5.4 -5.9. All the nanostructures exhibit good photocatalytic activity. One possible reason is that the vacancy of metal cations and O^{2-} anions in perovskite semiconductor structure results in the remarkable increase of oxygen adsorbed in the surface, which boosts the photocatalytic reaction. Perovskite oxide materials possessing bigger lattice distortion have higher photocatalytic activities by virtue of increasing the additional routes of trapping holes and decreasing the recombination rate of electron-hole pairs. Larger lattice distortion is probably another reason for the high photocatalytic activities of all perovskite compounds [12]. Meanwhile, the generated conduction band electrons (e^-) probably react with dissolved oxygen molecules to yield superoxide radical anions $(O^2)^-$. Which on protonation generate the hydroperoxy ($HO^\bullet_2$) radicals, producing the hydroxyl radical $OH^\bullet$ which acts as strong oxidizing agent to decompose the organic dye [4]. The relative reaction mechanism taking place is as follows

$$LaFeO_3 + h\nu \longrightarrow LaFeO_3\,(e^- + h^+)$$

$$LaFeO_3\,(e^-) + O_2 \longrightarrow LaFeO_3 + O_2^-$$

$$O_2^- + H_2O \longrightarrow HO^\bullet_2 + OH^-$$

$$HO^\bullet_2 + H_2O \longrightarrow H_2O_2 + OH^\bullet$$

$$H_2O_2 \longrightarrow 2OH^\bullet$$

$$OH^\bullet + RhB \longrightarrow \text{products}$$

In the present work, experiments were performed for nine different conditions *i.e.* (1) without catalysts, (2) dark, (3) Degussa P25 as catalyst and (4-9) $LaFeO_3$ nanocubes, flowers, dendrites, nanorods, microspheres, nanospheres as catalyst. Photocatalytic activity tests were conducted for enhanced degradation of RhB in aqueous solution under visible light irradiation. However, under dark condition without light illumination, the content of RhB remained constant for every measurement *i.e.* no change in absorption was noticed for all six $LaFeO_3$ nanostructures. Illumination in the absence of $LaFeO_3$ nanostructures did not result in the photocatalytic decolorization of RhB. Therefore, the presence of both illumination and $LaFeO_3$ samples were necessary for the efficient degradation of RhB. These results also suggest that decolorization of RhB aqueous solution

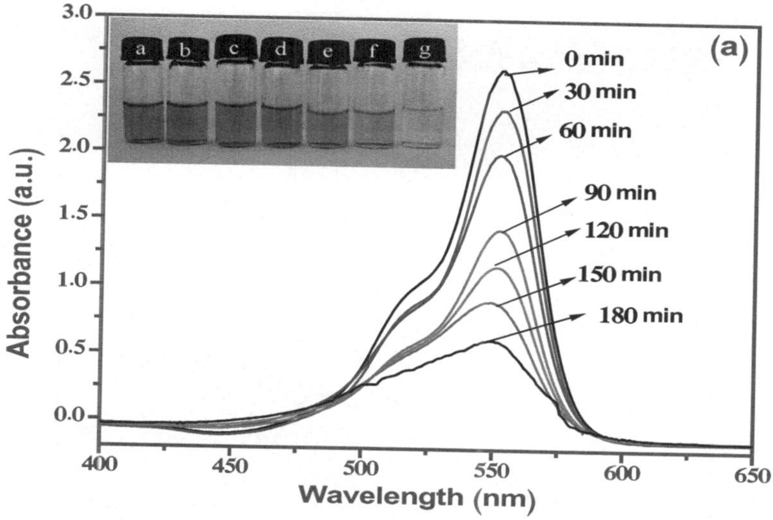

Figure 5.4. UV–Vis Time Dependent Absorbance Spectra during Photocatalytic Reaction of RhB by Nanocubes (inset) picture of analytical samples (3 ml) drawn from the reaction suspension every 30 min.

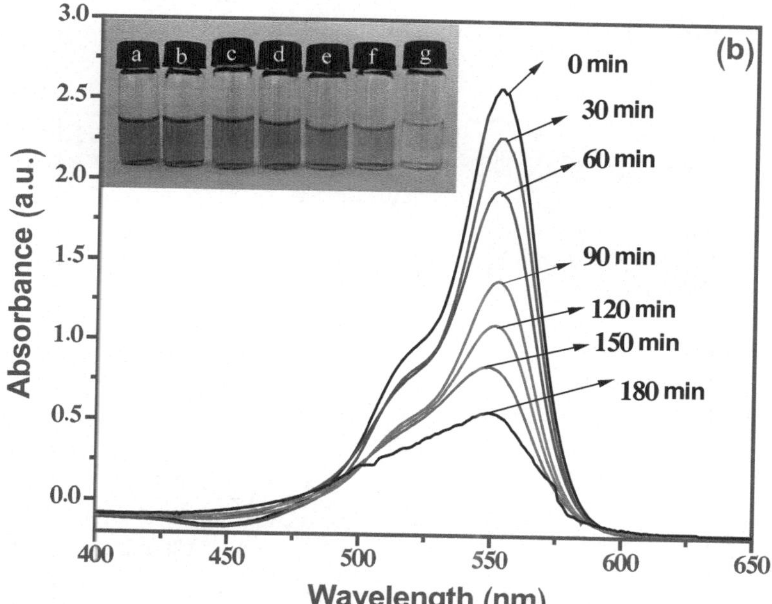

Figure 5.5. UV–Vis Time Dependent Absorbance Spectra during Photocatalytic Reaction of RhB by Microspheres, (Inset) Picture of Analytical Samples (3 ml) Drawn from the Reaction Suspension every 30 min.

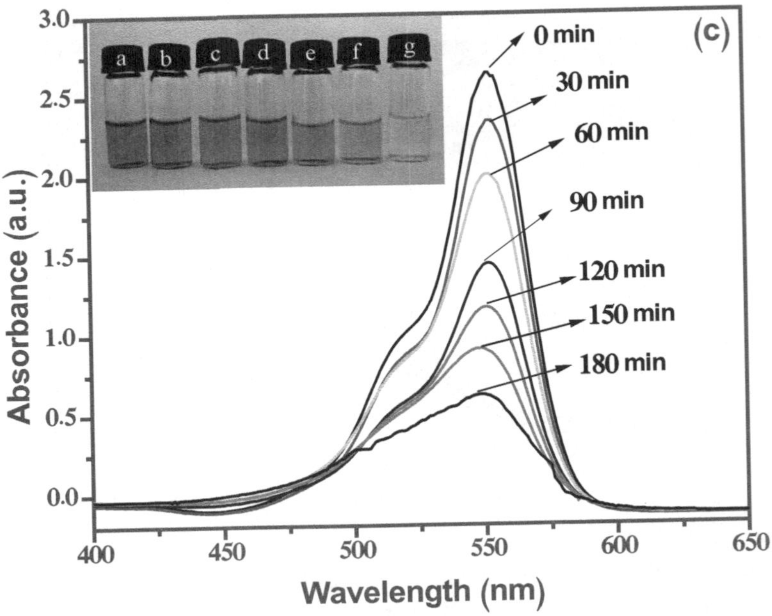

Figure 5.6. UV–Vis Time Dependent Absorbance Spectra during Photocatalytic Reaction of RhB by Nanospheres, (Inset) Picture of Analytical Samples (3 ml) Drawn from the Reaction Suspension every 30 min.

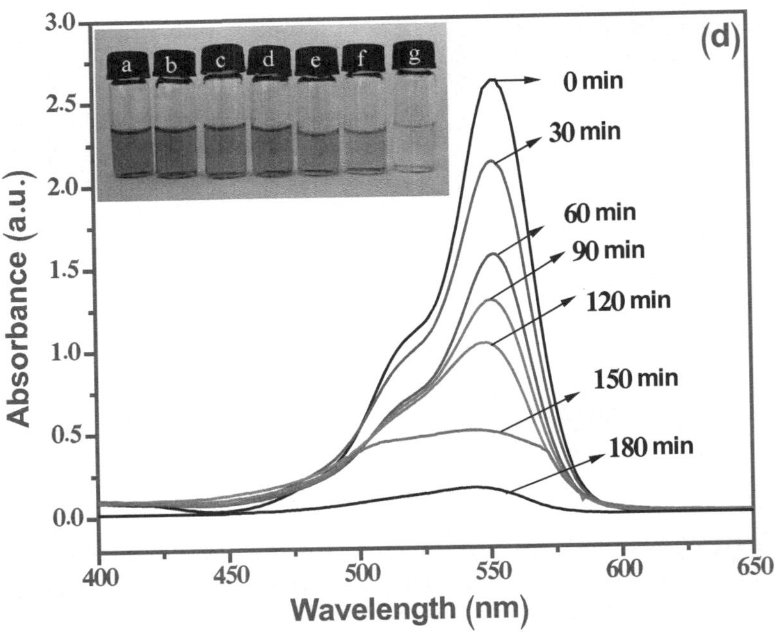

Figure 5.7. UV–Vis Time Dependent Absorbance Spectra during Photocatalytic Reaction of RhB by Flowers, (Inset) Picture of Analytical Samples (3 ml) Drawn from the Reaction Suspension every 30 min.

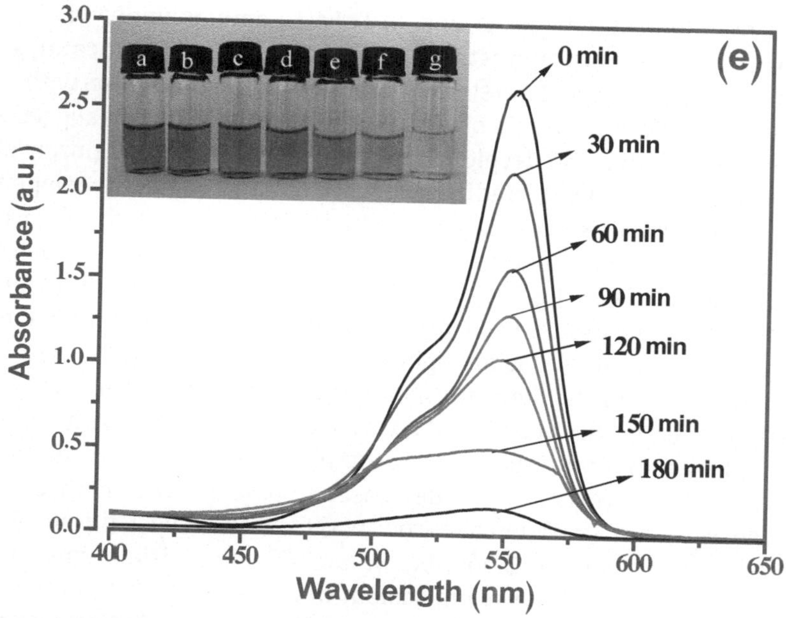

Figure 5.8. UV–Vis Time Dependent Absorbance Spectra during Photocatalytic Reaction of RhB by Dendrites, (Inset) Picture of Analytical Samples (3 ml) Drawn from the Reaction Suspension every 30 min.

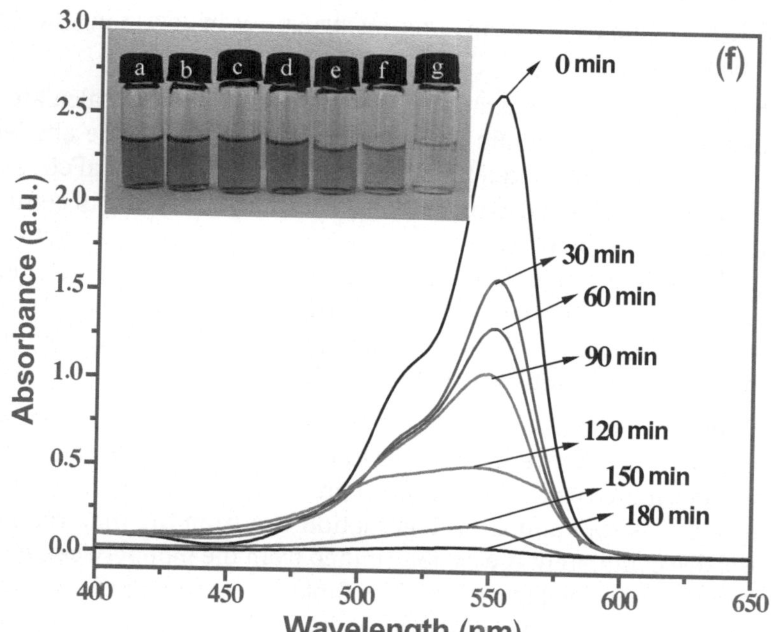

Figure 5.9. UV–Vis Time Dependent Absorbance Spectra during Photocatalytic Reaction of RhB by Nanorods, (Inset) Picture of Analytical Samples (3 ml) Drawn from the Reaction Suspension every 30 min.

was caused by photocatalytic reactions on $LaFeO_3$ samples under visible light illumination. When $LaFeO_3$ nanostructures were used as photocatalyst, the absorption maximum at $\lambda = 554$ nm steadily decreased upon irradiation with visible light without any obvious peak shift of the absorption maximum. This suggests that the complete degradation and decolourization of aqueous RhB was purely due to the catalyst under visible light irradiation. Figure 5.4, shows the time dependent absorbance spectrum of $LaFeO_3$ catalyst with nanocubes. It shows the absorbance maximum at 2.62 before irradiation with visible light and decreases to a value of 0.61 after 180 min of visible light exposure. The maximum occurs at 2.62 for both microspheres and nanospheres used as catalyst and it decreases to 0.52 and 0.41 respectively after 180 min of visible light exposure as shown in Figure 5.5 and 5.6. In the case of flowers and dendritic nanostructures, the maximum absorption peak of both samples occurs at 2.62 and it was dropped gradually with increasing the visible light irradiation time. After 180 min of visible light irradiation the absorption peak value weakened and decreased to 0.52 and 0.41 respectively as shown in Figure 5.7 and 5.8. When nanorods was used as catalyst, the absorbance maximum value decreased from 2.66 to 0.10, after 180 min of visible light exposure as shown in Figure 5.9. The intense pink colour of the starting RhB solution faded as irradiation time increased. The colour change of RhB solution indicates that RhB is de-ethylated and the destruction of its conjugated structure occurs under visible light irradiation. Prolonged exposure of visible light after 180 min did not exhibit any further decrease in the absorption maximum for all the $LaFeO_3$ nanostructures. The overall degradation trend of the different structures in degrading the dye RhB is shown in Figure 5.10.

The efficiency and rate constant of photodegradation were calculated with the value of absorption maximum and the degradation value. The efficiency of decolorization [13] was calculated for all the six morphologies of $LaFeO_3$ samples and is shown in Table 5.2. Furthermore, the photocatalytic activity of commercial TiO_2 powder (P25) and blank experiment were also tested. It was noticed that without any photocatalyst (blank) the degradation of RhB was extremely slow and only about 2.0 per cent. The commercial TiO_2 powder shows lower photocatalytic degradation for RhB under visible light irradiation, only 5.1 per cent within 180 min, TiO_2 absorption does not occur in the visible light region.

The decolourization efficiency of the different catalyst was calculated based on the initial and final absorbance value, using the formula in 5.11. The apparent rate constant was determined using the formula represented in 5.12, by considering the absorbance values, before and after degradation and from the time required for degrading the dye completely. It was determined from the plot between A_t/A_0 and irradiation time as shown in Figure 5.10. The Table 5.2 lists out the decolourisation efficiency and degradation rate of the different $LaFeO_3$ nanostructure catalysts prepared in the present study. From these experimental results presented in Table 5.2, it can be observed that there is a variation in the degradation rate of the different samples and the decolorization efficiency.

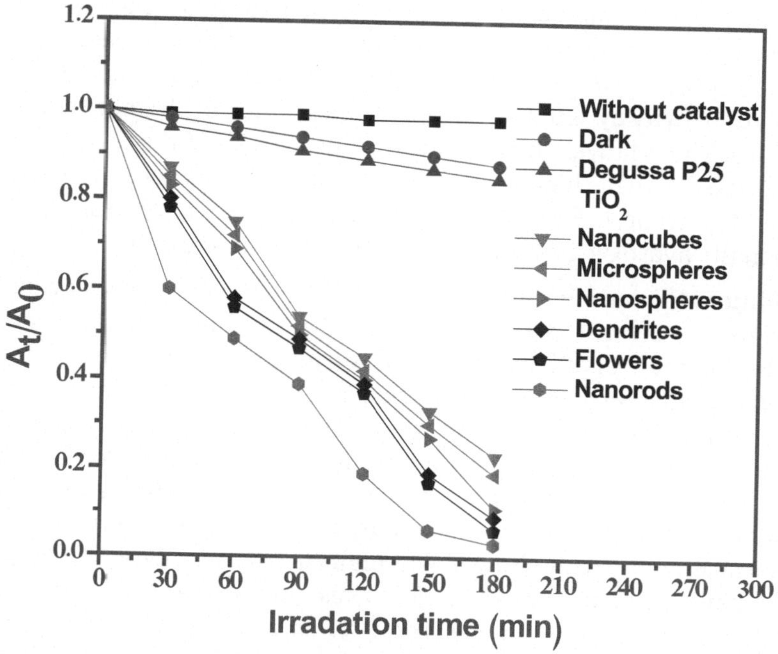

Figure 5.10. Degradation Trend of RhB as a Function of Irradiation Time in the Presence of different $LaFeO_3$ Nanostructures.

Table 5.2. The Decolourization Efficiency and the Apparent Rate Constant Values of the different $LaFeO_3$ Nanostructures Prepared in the Present Study

Catalyst	*Initial Absorbance (A_0)*	*Final Absorbance (A_t)*	*Time Required Domplete Degradation (t min)*	*Decolouri-sation Efficiency (per cent)*	*Rate Constant*
Nanocubes	2.62	0.61	180	76.71	0.7589
Microspheres	2.62	0.52	180	80.53	0.7926
Nanospheres	2.62	0.41	180	84.35	0.8505
Flowers	2.62	0.26	180	90.07	0.8762
Dendrites	2.62	0.18	180	93.12	0.8972
Nanorods	2.62	0.10	180	96.56	0.9124

The results suggest that all the $LaFeO_3$ nanostructure samples show higher photocatalytic activity for RhB degradation than Degussa P25 (TiO_2) in the visible light region. It is also evident that the $LaFeO_3$ nanorods show higher photocatalytic activity than other nanostructures such as nanocubes, flowers, dendrites, microspheres and nanospheres. From the degradation studies, it is observed that the visible light photocatalytic performance varies in the following order, Degussa P25 < nanocubes < microspheres < nanospheres < dendrites < flowers < nanorods. It has been reported that higher crystallization can usually lead to enhanced photocatalytic activity by suppressing the recombination of photogenerated holes and electrons. The relatively high photocatalytic activity of the nanorods can be closely associated with the high surface area of 84.4 $m^2 g^{-1}$ and pore size distribution of 1- 10 nm. Meanwhile, $LaFeO_3$ nanostructure samples have more adsorbed oxygen or oxygen in the hydroxyl group, as confirmed by the XPS results.

In addition, the specific surface area and the pore size distribution of the samples were already discussed in chapter 4. The specific surface area and pore size distribution confirms all the samples have a high surface area, which favoured the efficient degradation of the dye. Also, the pore size distribution analysis showed that all the samples prepared in the present study are mesoporous. Among the six different morphologies of $LaFeO_3$ evaluated for their photocatalytic efficiency, the nanocubes had a relatively less photocatalytic activity due to the lower specific surface area when compared to other morphologies. The nanorods have high specific surface area and it showed the highest photocatalytic efficiency. The other morphologies namely, flowers, dendrites, microspheres and nanospheres also followed the same trend. Hence, it can be ascertained that, specific surface area has a straightforward influence on the photocatalytic activity. It was already reported that, the presence of smaller and larger mesoporoes facilitates efficient transport of reactant molecules onto the active sites, enhancing the efficiency of photocatalysis [14]. In the present work, though all the samples were mesoporoes, the distribution of smaller and larger mesoporoes was found in nanocubes and nanorods. This might be the reason for high catalytic activity observed in the nanorods and nanocubes.

5.1.9 Conclusion

The present section, reports an efficient and facile route for the faster heterogeneous photocatalytic degradation of organic dye effluent RhB under visible light using $LaFeO_3$ nanostructures of different morphologies. The conventional drawback of slow decomposition rate during photocatalytic degradation has been greatly avoided by the use of nanostructures with superior surface properties. It has been qualitatively observed that, even a small change in the surface pore distribution can drastically change the degradation rate of the nanostructures. However, photoconversion activity is a function of the catalytic properties, the organic species to be degraded and the experimental condition. So, a catalyst having the highest activity for one compound might not necessarily be the best catalyst for the destruction of other organic species. Hence for this technology to be successful and reach widespread application, significant advancement in the assessment of reaction intermediates is still required. The present study thus offers a new insight into environmental benign and novel photocatalytic absorbents for efficient photocatalytic degradation of effluents.

PART- II

5.2 Electrocatalytic Performance of different Morphologies of $LaFeO_3$ Nanostructures

5.2.1 Introduction

Nanotechnology has recently become one of the most exciting forefront fields in analytical chemistry. A wide variety of nanoscale materials of different sizes, shapes and compositions at 1-100 nm scale, are now available. Perovskite oxide nanoparticles were applied in this work for designing novel sensing systems and it enhances the performance of bio-analytical assays and improving the visualisation of bio-interactions occurring on sensing surfaces. Many types of nanoparticles with different sizes and compositions are now available, which facilitate their application in electroanalysis, bringing important advantages like (a) their immobilization on electrode surface generates a roughened conductive high-surface area interface that enables the sensitive electrochemical detection of molecular and biomolecular analytes; (b) nanoparticles act as effective labels for the amplified electrochemical analysis of the respective analytes; (c) the conductivity properties of metal nanoparticles enable the design of biomaterial architectures with pre-designed and controlled electrochemical functions. Electrochemical sensors offer elegant routes for interfacing at the molecular level, chemical or biological recognition events and electronic signal-transduction processes. In addition, electrochemical devices are uniquely qualified for meeting the size, cost, low-volume, and power requirements of decentralized testing and indicate great promise for a wide range of biomedical or environmental applications.

5.2.2 Modification of Electrode Surfaces with Nanomaterials

Modified electrodes represent a modern approach to electrode systems. These electrodes rely on the placement of a reagent onto the surface, so that the electrode thereafter displays the chemical, electrochemical, optical and other properties of the immobilized materials or molecules. Thus, one selects immobilized materials or molecules on the basis of known and desired properties, to be, for instance, fast outer-sphere electron transfer agents, chiral centres, electron transfer mediator-catalysts for a valuable substrate reaction, functionalities which scavenge trace molecules or ions from solutions for analysis in preconcentrated form, photosensitizers for a semiconductor electrode corrosion inhibitors and so on. This means that the electrode can be tailor-made to exhibit these properties. Such deliberate alteration of electrode surfaces can thus meet the needs of many electroanalytical problems, and may form the basis for new analytical applications and different sensing devices. Such surface functionalization of electrodes with molecular reagents or materials finds many applications, including clinical, industrial and environmental monitoring. Many other important analytical applications like acceleration of electron transfer reactions, preferential accumulation, or selective membrane permeation should also benefit from the rational design of electrode surfaces.

Such steps can impart higher selectivity, sensitivity, or stability on electrochemical devices. These analytical applications and improvements have been extensively dealt with in literature.

5.2.3 Electroanalytical Techniques

Currently, there are several techniques available to investigate the application of nanomaterials as chemical sensors. Among those, electroanalytical techniques offer tremendous promise for scaling down analytical systems, with features that include high sensitivity, inherent miniaturization, low cost, low-power requirements, and high compatibility with advanced micromachining and microfabrication technologies. Whereas the progress in electroanalytical methods in 1960s-1970s was mainly based on the development of new electrochemical techniques such as different kinds of alternating current voltammetry, pulse-voltammetry, and stripping voltammetry, further developments in electroanalytical chemistry during the 1980s-1990s originated from advances accomplished by the chemical modification and functionalization of electrodes for enhanced sensitivity and selectivity of electroanalysis [15]. Bioelectroanalytical chemistry also received tremendous impact from the development of the methods of electrode modifications. Reversible electrochemistry of redox proteins (*e.g.*, cytochromes) as promoter modified electrodes [16] and electrical wiring of redox enzymes with immobilized electron-relays became possible.

Electroanalytical techniques are concerned with the interplay between electricity and chemistry, namely, the measurements of electrical quantities, such as current, potential, or charge and their relationship to chemical parameters. Such use of electrical measurements for analytical purposes has found a vast range of applications, including environmental monitoring, industrial quality control, or biomedical analysis. Indeed, electrochemical probes are receiving a major share of the attention in the development of chemical sensors. In contrast to many chemical measurements, which involve homogeneous bulk solutions, electrochemical processes take place at the electrode–solution interface. The distinction between various electroanalytical techniques reflects the type of electrical signal used for the quantitation. The two principal types of electroanalytical measurements are potentiometric and potentiostatic. Both types require at least two electrodes (conductors) and a contacting sample (electrolyte) solution, which constitute the electrochemical cell. The electrode surface is thus a junction between an ionic conductor and an electronic conductor. One of the two electrodes responds to the target analyte(s) and is thus termed the indicator (or working) electrode. The second one, termed the reference electrode, is of constant potential (*i.e.*, independent of the properties of the solution). In a three-electrode cell, a counter electrode is used only to make an electrical connection to the electrolyte so that a current can be applied to the working electrode. Controlled-potential (potentiostatic) techniques deal with the study of charge transfer processes at the electrode–solution interface, and are based on dynamic (non-zero-current) situations. Here, the electrode potential is being used to derive an electron transfer reaction and the resultant current is measured. Such a controllable parameter can be viewed as "electron pressure," which forces the chemical species to gain or lose an electron (reduction or oxidation, respectively). Accordingly, the resulting current reflects the rate at which electrons move across the electrode-solution interface. Potentiostatic techniques can thus measure any chemical species that is electroactive *i.e.* that can be made to reduce or oxidize. Knowledge of the reactivity of functional group in a given compound can be used to predict its electroactivity. Non-electroactive compounds may also be detected in connection with indirect or derivatization procedures. The advantages of controlled-potential techniques include high sensitivity, selectivity toward electroactive species, a wide linear range, portable and low-cost instrumentation, speciation capability, and a wide range of electrodes that allow assays of unusual environments. Extremely low (nanomolar) detection limits can be achieved with very small (5–20-µL) sample volumes, thus allowing the determination of analyte amounts ranging from 10^{-13} to 10^{-15} M on a routine basis. Improved selectivity may be achieved.

5.2.4 Cyclic Voltammetry (CV) and Linear Sweep Voltammetry (LSV)

Cyclic voltammetry (CV) and linear sweep voltammetry (LSV) are the most widely used voltammetric techniques for studying redox reactions of both organic and inorganic compounds because they are unmatched in their ability to provide information on the steps involved in electrochemical processes with only a modest expenditure of time and effort in the acquisition and interpretation of data. These electroanalytical methods require simple and inexpensive instrumentation and provide information not only on the electrochemical quantities typical of a redox process, but also allow investigations of chemical reactions coupled with charge transfer steps. This is because the electrode can be used as a tool for producing reactive species in a small solution layer surrounding its surface and at the same time to monitor chemical reactions involving these species. Moreover, since the relevant responses can be obtained within a few milliseconds after stimulation of the electrode, they may be used for studying mechanisms involving very fast reactions, thus allowing detection of short-lived transient intermediates. In both LSV and CV, a small (<0.1 cm^2) stationary working electrode is dipped in an unstirred solution containing excess of supporting electrolyte to repress migration of charged reactants and products, so that any transfer of electroactive species to and from the electrode surface can occur only through diffusion. In LSV, the potential of the working electrode is changed linearly with time, starting from a potential where no electrode reaction occurs and moving to potential where reduction (more negative values) or oxidation (more positive values) of the investigated analyte takes place. The potential is measured between the reference electrode and the working electrode and the current is measured between the working electrode and the counter electrode. Oxidation or reduction of species is registered as a peak or trough in the current signal at the potential at which the species begins to be oxidized or reduced. The current is measured throughout the experiment and the resulting current–potential curve (voltammogram) is plotted.

5.2.5 Amperometric

Amperometric is an electroanalytical technique that involves the application of a constant reducing or oxidizing potential to an indicator (working) electrode and the subsequent measurement of the resulting steady-state current. Owing to the time dependence of the resulting anodic or cathodic current, the reliable analytical utilization of amperometry requires measurement at a fixed time interval. The magnitude of the measured current is dependent on the concentration of the reduced or oxidized substance, and hence this method has been widely used for the analysis of various substances in samples ranging from water to radioactive materials. Also, amperometric sensors, such as the dissolved oxygen probe and various amperometric biosensors, are widely used for clinical, environmental, and industrial monitoring [17]. Furthermore, amperometric detectors have gained considerable use since the 1970s in high-performance liquid chromatographic determination of various substances and in flow injection analysis.

5.2.6 $LaFeO_3$ Electrocatalysts for Dopamine Biosensing

The desire and sense important physiological analytes with high sensitivity has fuelled the recent marriage between nanotechnology and biosensors. Nowadays, advances in biosensors based on nanostructured perovskite oxides are of practical importance in biological science, environmental science, neurochemistry and biomedical chemistry [18]. L-Dopa, the precursor of dopamine (DA) is a chemical messenger produced within the nerve cell that is essential for the neurotransmission of the nerve impulse and hence the dysfunction in the central nervous dopamine system can lead to diseases such as Parkinson's and Schizophrenia. *In-vitro*, dopamine is a powerful toxin to the culture of neurons and may also be toxic *in-vivo* according to some animal studies [19]. Therefore, the determination of DA has attracted much attention of researchers. Furthermore, some bio-molecules such as uric acid (UA) and ascorbic acid (AA) have similar oxidation potential as that of dopamine and it shows serious interference during the determination of dopamine [20]. To overcome these problems and also considering cost evaluation, huge attention have been turned towards perovskite oxide based nanomaterials modified glassy carbon electrode as alternatives. However, to the best of our knowledge, the cost effective novel different morphologies of $LaFeO_3$ nanostructures for dopamine biosensing application has not been explored in open literature.

5.2.7 Experimental Procedure

(a) Experimental Setup

All the electroanalytical methods performed in this research used a reference electrode, working electrode, and counter electrode which in combination are sometimes referred to as a three-electrode setup. The potential is measured between the reference electrode and the working electrode and the current is measured between the working electrode and the counter electrode. An electrolyte is usually added to the test solution to ensure sufficient conductivity. The combination of the solvent, electrolyte and specific working electrode material determines the range of the potential. The experimental setup for the electroanalytical methods is shown in Figure 5.11. Electric signals were measured and recorded with potentiostat - galvanostat electrochemical workstation (EG and G Instrument model 6310). The prepared solutions were deserted by purging with nitrogen gas for 10 min prior to the electrochemical experiments. The electrochemical measurements were performed in a conventional two compartment three electrode cell with mirror polished 0.07 cm^{-2} glassy carbon as the working electrode, Pt wire as the counter electrode and 3M KCl Ag/AgCl as the reference electrode. All the measurements were carried out in phosphate buffer solution (pH = 7.2) under nitrogen atmosphere at room temperature.

Figure 5.11. Photograph of Electrocatalytic Reaction Setup.

(b) Electrode Preparation

The bare glassy carbon electrode was polished to a mirror like surface with 0.05 μM alumina powder and sonicated in double distilled water for 10 min. 100 mg of $LaFeO_3$ nanoparticles was dispersed in 10 mL anhydrous alcohol and ultrasonicated for 30 min, 10 μL of alcoholic dispersed $LaFeO_3$ from the above solution was dropped onto the glassy carbon surface and air dried at ambient temperature.

5.2.8 Results and Discussion

5.2.8.1 Electrocatalytic Properties of different Morphologies of $LaFeO_3$ Nanostructures

From the technological point of view, it is very important to develop biosensors with high sensitivity, ease of operation, and high efficiency for dopamine detection.

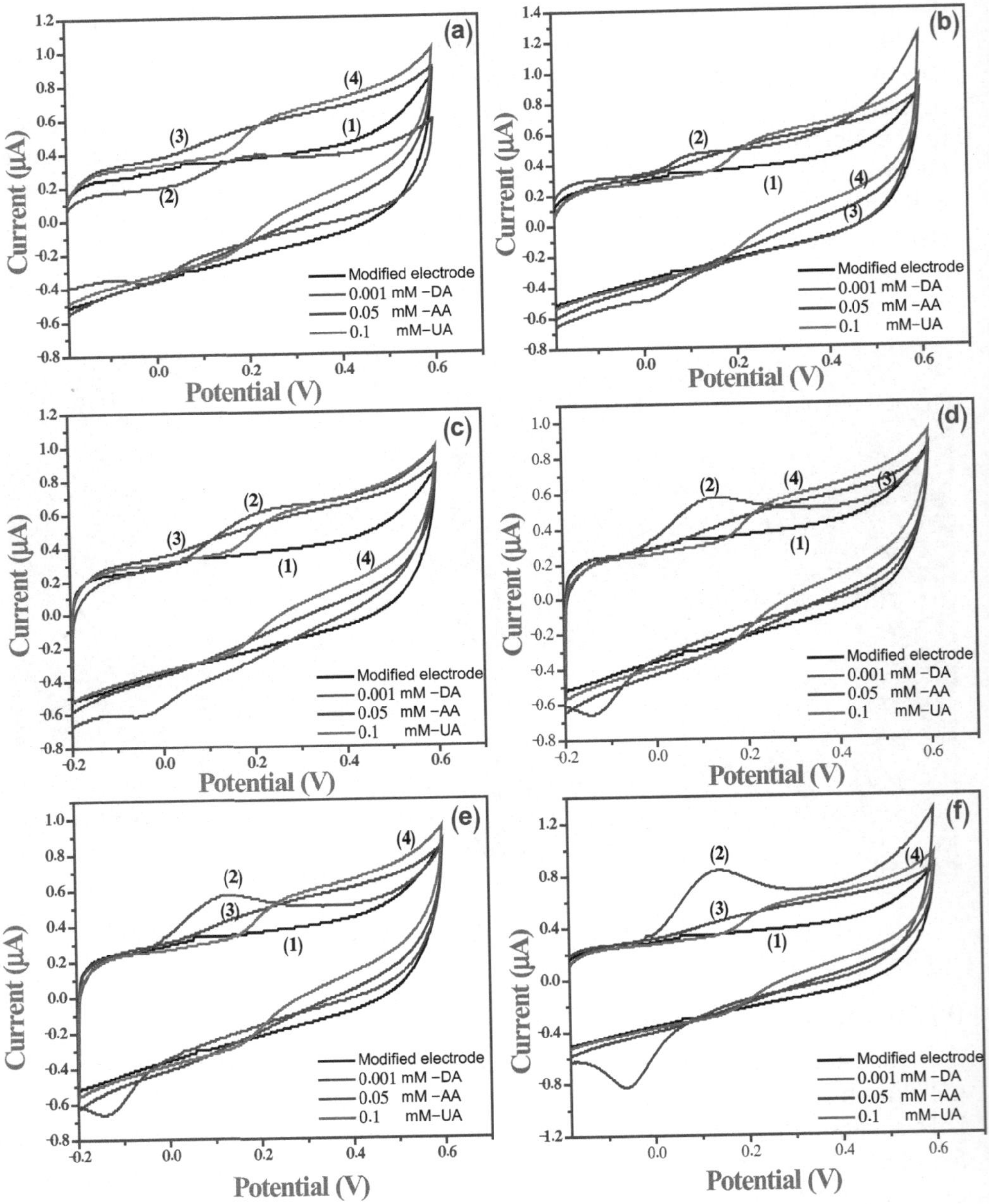

Figure 5.12. Cyclic Voltammograms Recorded for different $LaFeO_3$ Nanostructures (a) Nanocubes, (b) Microspheres, (c) Nanospheres, (d) Flowers, (e) Dendrites and (f) Nanorods Modified Electrode in PBS (10 mM, pH 7.2) at 50 mVs^{-1} for (a) Low Concentration of DA, High Concentration of AA, and UA.

Recently, trends in the research and development of novel nanostructures have increasingly emphasized the application in biosensing. Although perovskite oxide nanostructures have wide and significant applications in many fields, only few reports have been focused for bio-molecular detection. Herein, we report different morphologies of $LaFeO_3$ nanostructures modified glassy carbon electrode for sensing dopamine (DA), ascorbic acid (AA) and uric acid (UA).

Figure 5.12 shows cyclic voltammetric (CV) curve for different morphologies of $LaFeO_3$ nanostructures modified glassy carbon electrode (GCE) and the modified electrode with DA, AA and UA recorded in 10 mM PBS (pH = 7.2) solution. In order to avoid the environmental effects, we measured CV curves for the solution saturated with N_2 gas.

It can be observed from Figure 5.12 (a-f), that DA, AA and UA show no redox peaks for the modified GCE. It is clear that the oxidation peak potentials of DA, AA, and UA are very close to each other for $LaFeO_3$ nanostructures modified electrode and the electrochemical reactions are quasireversible or irreversible. For all $LaFeO_3$ nanostructures 1 μM DA (red colour, Figure 5.12 a-f), the cathodic and anodic peaks appear in the range, -135 to - 145 mV and 125 to 135 mV respectively, and the peak potential separation (ΔE) is about 250 to 260 mV at modified GCE, the oxidation peak results from the dopamine oxidation to dopaminequinone, while the reduction peak is caused by reducing dopaminequinone to dopamine. For 50 μM AA (blue colour, Figure 5.12 a-f) and 100 μM UA (green colour, Figures 5.12 a-f), the electrochemical reactions are irreversible and the signals show broader oxidation peaks with the peak potentials at 134 to 145 mV (AA) and 223 to 233 mV (UA). These indicate the sluggish electron transfer kinetics at the $LaFeO_3$ nanostructures, which may be related to the nonconductive components and the electrode fouling caused by the deposition of AA, UA and their oxidation products on the electrode surface [21]. The oxidation for DA, AA and UA appears in the range 125 to 135, 134 to 144 and 223 to 224 mV, respectively. The peak potential of DA-AA, UA-AA and UA-DA with low oxidation peak separation of 9 to 19, 89 to 99 and 98 to 108 mV respectively were noticed all the samples. These separations are not sufficient enough for simultaneous determination of these species. It is observed that the concentration of DA is as low as 50 times AA and 100 times UA. The excellent selectivity of the $LaFeO_3$ nanostructures modified electrode for DA can be explained as follows; the metal surface of $LaFeO_3$ dendrites modified electrode was negatively charged, AA and UA exist as anions and DA exist as cations. The negative surface of the metal $LaFeO_3$ nanostructures attracts the cation DA and simultaneously repels AA and UA anions. So AA and UA cannot enter the $LaFeO_3$ nanostructures surface to the same extent as DA, which is clearly evident from the redox peak at -135 to -145 mV. Therefore, the $LaFeO_3$ nanostructures exhibit strong electrocatalytic activity in response to dopamine reduction and improve the electron transfer between dopamine and the morphology dependent $LaFeO_3$ nanostructures modified GCE.

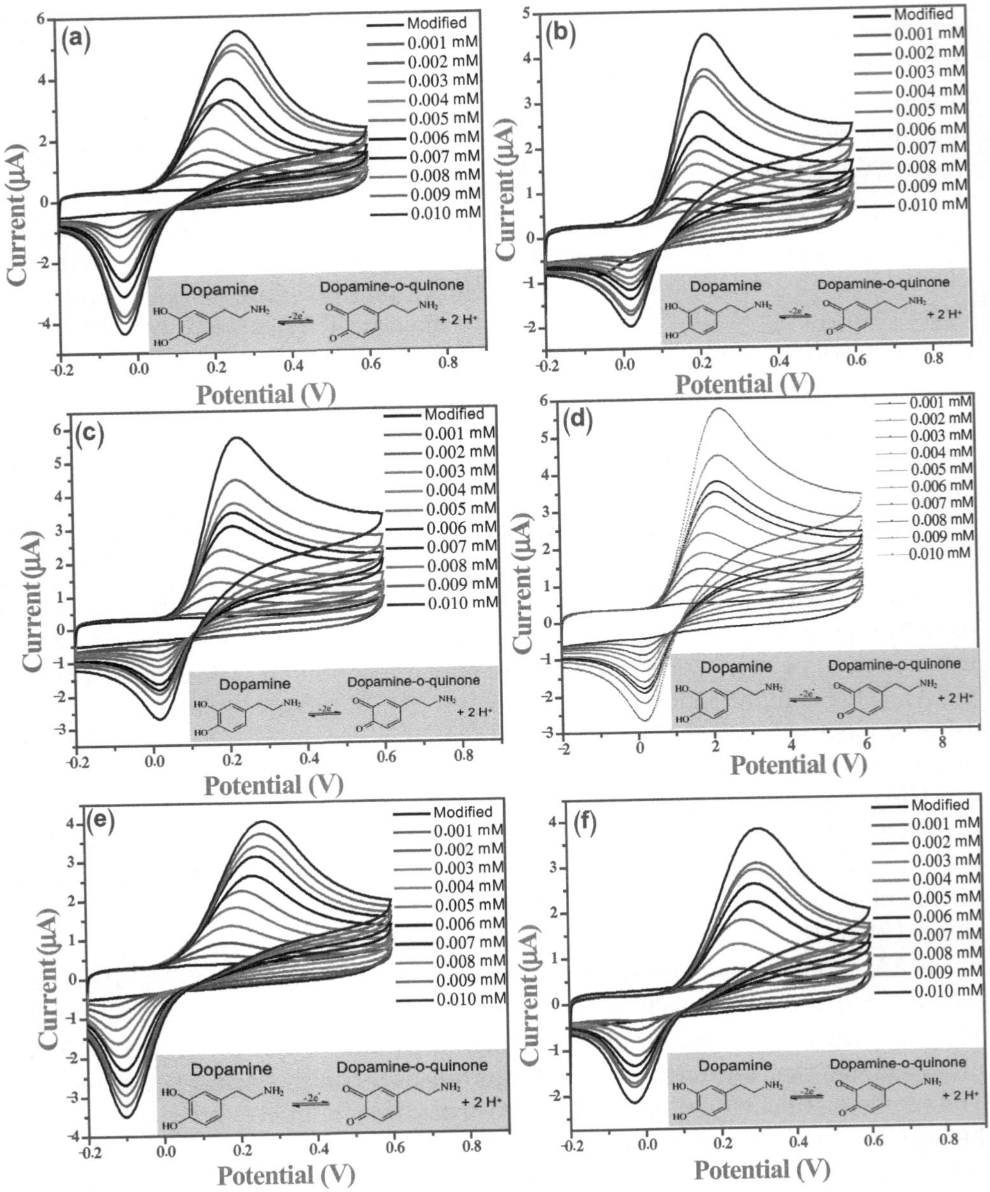

Figure 5.13. Cyclic Voltammograms Recorded for different Morphologies of $LaFeO_3$ Nanostructures (a) Nanocubes, (b) Microspheres, (c) Nanospheres, (d) Flowers, (e) Dendrites and (f) Nanorods Modified Electrode in PBS (10 mM, pH 7.2) with different Concentration of DA (1 μM to 10 μM).

The dopamine electrocatalytic ability of different morphologies $LaFeO_3$ nanostructures modified electrode at different concentration was obtained for further investigations. Figure 5.13 (a-f) shows the electrocatalytic properties of

$LaFeO_3$ nanostructures towards the reduction of dopamine in 10 mM PBS (pH = 7.2) with a scan rate of 50 mV/S and different dopamine concentrations (1 μM to 10 μM). It can be observed that the dopamine peak current increases with the increase in dopamine concentration.

The oxidation current has a linear relationship with concentration of DA, with a correlation coefficient of 0.9920 to 0.9986 (data not shown). Figure 5.14 (a-f), shows the linear sweep voltammograms (LSVs) obtained for DA in the concentration range of 0-100 μM with the $LaFeO_3$ nanostructures modified electrode. The oxidation current of DA got enhanced with the increase in DA concentration (each step with the increment of 5 μM). A good linearity between the anodic peak current and the reaction of the electrode with dopamine was observed due to diffusion controlled process. From the plot of catalytic current Vs the concentration of DA (inset) it was observed that the catalytic current for the oxidation of DA at modified $LaFeO_3$ dendrites was proportional to the concentration of DA in the range 1.7×10^{-8} to 4.1×10^{-7} M and was represented by a linear equation [22] with a correlation coefficient of 0.9921 to 0.9982. The sensitivity and the lower limit were found to be 89 to 65 nM and 1.5×10^{-7} to 8.9×10^{-8} M, respectively.

According to the above discussions, the redox steps involved during the cyclic voltammetric response of $LaFeO_3$ nanostructures modified electrode for dopamine sensing has been illustrated schematically in Figure 5.15. The electrocatalytic mechanism involves the electrochemical oxidation of Fe (III) producing Fe (IV) complex on the surface of electrode, followed by the electron transfer of dopamine and consequently regeneration of Fe (III) in the complex. The oxidation of dopamine to dopaminequinone by liberating two hydrogen can be catalyzed by the Fe (IV)/(III) redox couple in the alkaline medium, also confirmed from the oxidation and reduction peaks in Figure 5.13.The following electrochemical reactions are considered as the basis for the fabrication of neurotransmitter compound sensor for electrochemical detection of dopamine during the electrooxidation mechanism.

$$LaFeO_3 - Fe\,(III) + OH^-_{Sol} \Longleftrightarrow LaFeO_3(OH^-) - Fe\,(IV) + e^- \quad (1)$$

$$LaFeO_3(OH^-) - Fe\,(IV) + dopamine \Longrightarrow LaFeO_3 - Fe\,(III) + dopaminequinone \quad (2)$$

The $LaFeO_3$ nanostructures modified electrode exhibited high electrocatalytic activity towards dopamine oxidation, which improves the reversibility and enhances the electron transfer kinetics. The incorporation of Fe (IV) complex in the $LaFeO_3$ dendritic structures helped to improve the dopamine electrocatalytic activity of the modified electrode.

5.2.8.2 Amperometric Determination of DA

Furthermore, the amperometric study was performed to examine the sensitivity of different morphologies of $LaFeO_3$ nanostructures modified electrode towards the detection of dopamine in a homogeneously stirred 10 mM PB solution (pH 7.2) as shown in Figure 5.16 (a-f). The $LaFeO_3$ nanostructures modified electrode

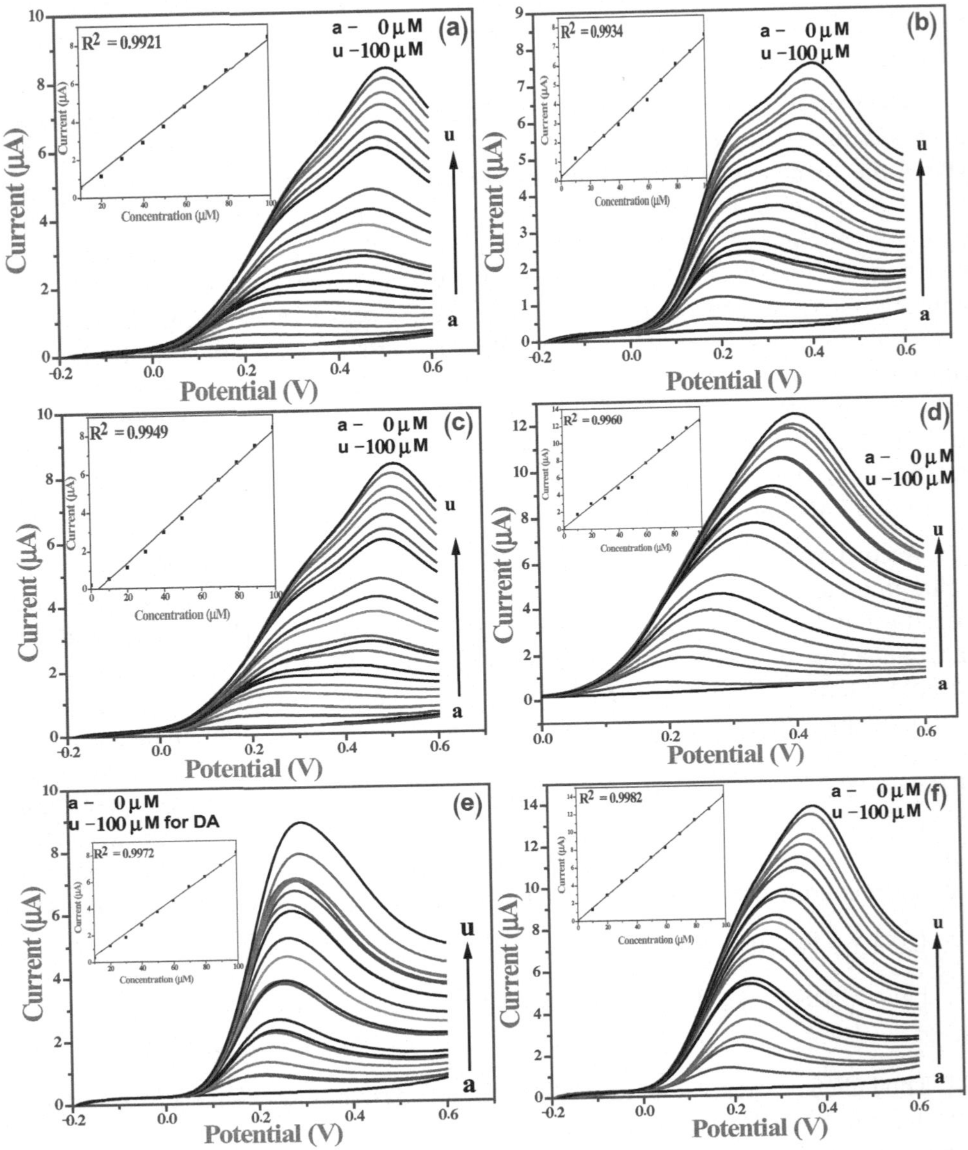

Figure 5.14. LSVs Obtained for DA Concentration Range from 0 to 100 μM for different Morphologies of $LaFeO_3$ Nanostructures (a) Nanocubes, (b) Microspheres, (c) Nanospheres, (d) Flowers, (e) Dendrites and (f) Nanorods Modified Electrode in PBS (10 mM, pH 7.2) with different Concentration of DA (1 μM to 10 μM) (Inset), Corresponding Plot of Catalytic Current Vs. the Concentration of DA.

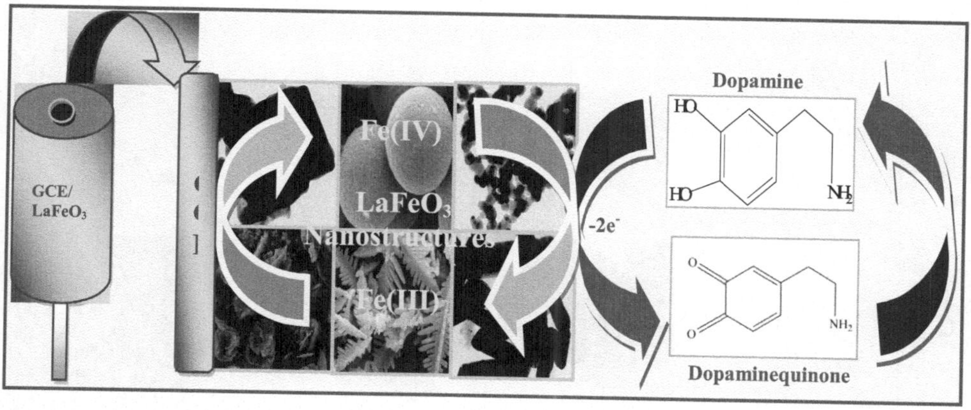

Figure 5.15. Schematic Illustration of Electrocatalytic Dopamine Oxidation at different $LaFeO_3$ Nanostructure Modified GCE.

shows the current response for the addition of each 100 nM DA. The current response increases and the steady state current response were attained within 5s for further additions of 100 nM DA in each step with a sample interval of 180 s. The dependence of response current on the concentration of DA was linear from 100 nM to 1500 nM with a correlation coefficient in the range 0.9924 to 0.9984 as shown in Figure 5.16 (a-f) (inset).The amperometric i–t curve for each addition of 100 nM DA showed a linear current increase without noise. It is worthy to compare the detection limit and wide linear range concentration of DA obtained in the present study with the reported different nanostructures modified electrodes. All $LaFeO_3$ nanostructures modified electrode showed the lowest detection limit for dopamine with a wide range of concentration as shown in Table 5.3. This value is very much comparable to the literature values in Table 5.4. These results confirmed that the $LaFeO_3$ nanostructures modified electrode has high selectivity and good sensitivity towards DA.

5.2.8.3 Reproducibility and Stability of different $LaFeO_3$ Nanostructure Electrode

The reproducibility of different morphologies $LaFeO_3$ nanostructures modified electrode was evaluated at a given concentration in the linear range by cyclic voltammetric measurements. The relative standard deviations of the nanostructured $LaFeO_3$ biosensors at 1 μM response for 7 successive measurements are 3.5 to 3.9 per cent, indicates a good reproducibility. The stability of biosensors was studied by comparing the cyclic voltammetric peak currents of the $LaFeO_3$ dendritic nanostructures with the time intervals of 4 h.

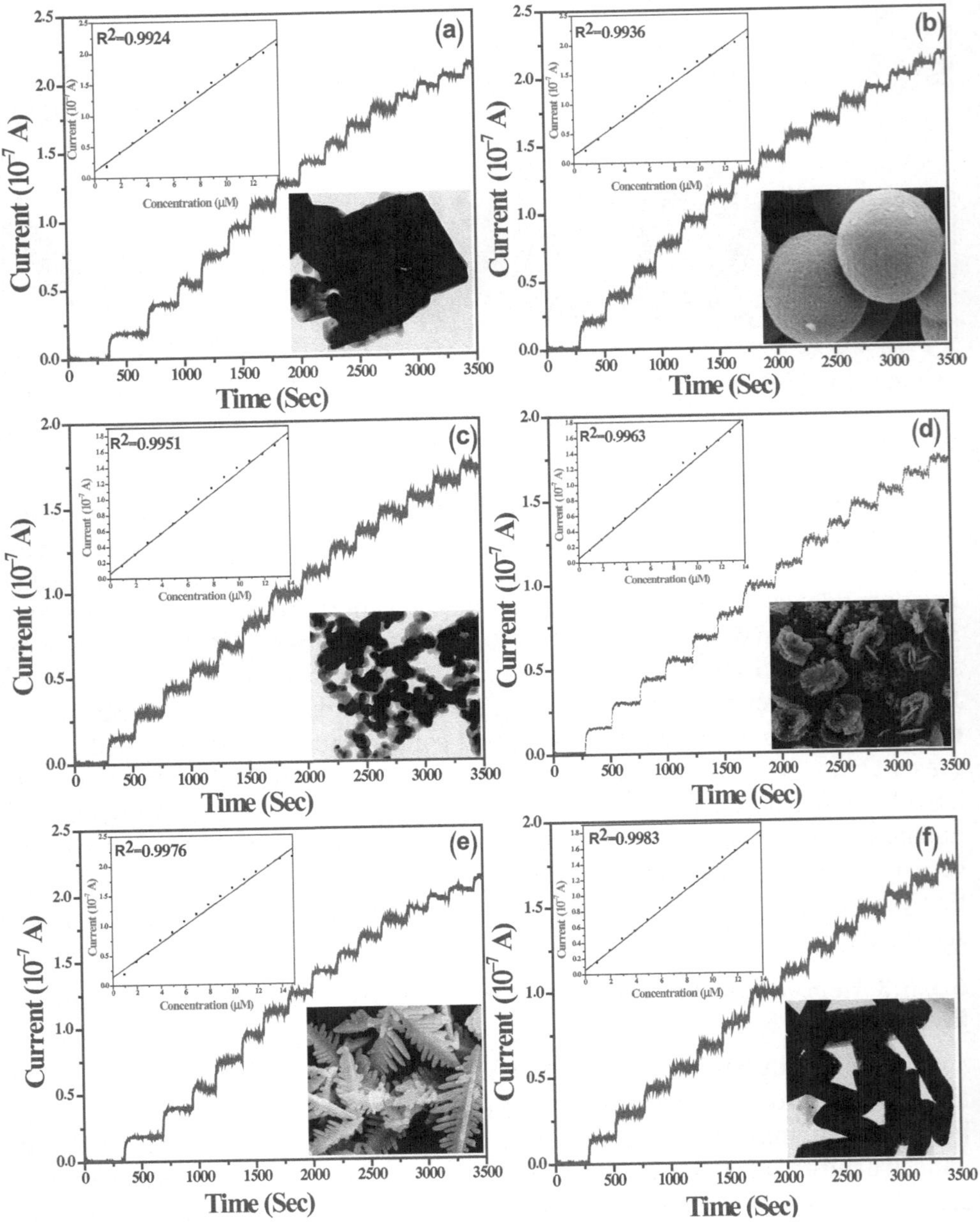

Figure 5.16. Amperometric i-t Curve for the Determination of DA with different Morphologies $LaFeO_3$ Nanostructures Modified Electrode in 10 mM PBS 7.2. Each addition increases the concentration of 100 nM at regular interval 180s, (inset top) plot of the concentration of DA Vs current and (inset down) corresponding nanostructures.

Table 5.3 Performance of different $LaFeO_3$ Nanostructures Based on Dopamine Biosensing

Nanostructures	*pH*	*Linear Range (M)*	*Detecting Limit (nM)*
Nanocubes	7.2	8.7×10^{-7} to 5.6×10^{-6}	89
Microspheres	7.2	9.5×10^{-7} to 6.2×10^{-6}	85
Nanospheres	7.2	1.7×10^{-8} to 4.3×10^{-6}	82
Flowers	7.2	4.5×10^{-8} to 6.5×10^{-6}	79
Dendrites	7.2	6.3×10^{-8} to 3.8×10^{-6}	72
Nanorods	7.2	8.2×10^{-8} to 1.6×10^{-6}	64

The decrease in the cathodic peak currents of $LaFeO_3$ biosensors immersed in PBS for 4 h was less than 3.7 per cent, indicates that this possess a good stability. Moreover, this biosensor can retain 98.2 per cent of their initial response after one week of aging effect, suggests good long-term stability. The excellent reproducibility and stability of all $LaFeO_3$ nanostructure biosensor can be contributed to the unique shape, and good biocompatibility.

Table 5.4. Performance of the $LaFeO_3$ Nanostructures Based Dopamine Biosensing in Comparison with Reported other Nanomaterials Sensor

Electrodes	*pH*	*Linear Range (M)*	*Detecting Limit (M)*	*Ref*
GNP – MEA - NIHCF	7.0	8.2×10^{-7} to 2.5×10^{-3}	53×10^{-8}	23
Polycarbazole electrode	7.0	1.0×10^{-3} to 1.0×10^{-2}	68×10^{-6}	24,25
Oxavandium thin film electrode	6.0	1.0×10^{-6} to 1.0×10^{-4}	80×10^{-8}	26
PEDOT-SWNT	7.0	2.0×10^{-5} to 1.0×10^{-1}	10×10^{-8}	27
WO_3. H_2O-GCE	7.2	1.0×10^{-7} to 1.0×10^{-6}	10×10^{-8}	28
$LaFeO_3$ materials	7.2	1.5×10^{-7} to 8.0×10^{-4}	30×10^{-8}	29
$LaFeO_3$ Nanorods	7.2	8.2×10^{-8} to 1.6×10^{-6}	64×10^{-9}	This work

The results suggest that all the $LaFeO_3$ nanostructure samples show higher electrocatalytic activity for dopamine biosensing than other nanostructure materials. It is also evident that the nanocubes, which had a relative lower specific surface area (40.7 $m^2 g^{-1}$) when compared to other morphologies, showed decreased electrocatalytic activity. On other hand, the $LaFeO_3$ nanorods show higher electrocatalytic activity than other nanostructures such as nanocubes, flowers, dendrites, microspheres and nanospheres. From the amperometric studies, it was observed that the detection limit performance varies in the following order, nanocubes < microspheres < nanospheres < dendrites < flowers < nanorods.

It has been reported that higher crystallization can usually lead to enhanced electrocatalytic activity by suppressing unusual electrochemical oxidation of Fe (III) producing Fe (IV) complex on the surface of electrode. The relatively high electrocatalytic activity of the nanorods can be closely associated with the high surface area of 84.4 m^2 g^{-1}. Meanwhile, $LaFeO_3$ nanostructure samples have more adsorbed oxygen or oxygen in the hydroxyl group, as confirmed by the XPS results.

5.2.9 Conclusion

In summary, the different morphologies of $LaFeO_3$ nanostructures were used to modify the electrode for dopamine detection limit at low range of 85 to 62 nM due to its higher surface area range in the 40.7 to 84.4 m^2g^{-1} and unusual electrochemical oxidation of Fe(III) producing Fe(IV) complex on the surface of electrode. The detection limit in the present study is one order of magnitude higher than the reported value of 300 nM for the same material. Also, the present sensor effectively avoids the interference of ascorbic acid and uric acid, with high selectivity and sensitivity for dopamine. Our results confirmed that the different morphology of $LaFeO_3$ nanostructures exhibits higher dopamine biosensing behavior compared to other reported materials. Most importantly, the obtained $LaFeO_3$ nanostructure with high surface area exhibits high electrocatalytic ability (selectivity, sensitivity, reproducibility and stability) for dopamine reduction when compared to that of other nanomaterials and found to be good candidate for application as dopamine biosensor in analytical detection. It is believe that $LaFeO_3$ nanostructures, a financially viable perovskite oxide could be a promising material with a great potential for application in many fields.

References

[1] T. Masciangioli, W. X. Zhang, *Environ, Sci. Technol*, **37** (2003) 102A.

[2] A. Mills,S. L. Hunte, *J. Photochem. Photobiol.*, **108** (1997) 1.

[3] A. L. Linsebigler, G. Lu, J. T. Yates, *Chem, Rev.*, **95** (1995) 735.

[4] H. Hattori, *Chem, Rev.*, **95** (1995) 537.

[5] M. R. Hoffmann, S. T. Martin, W. Choi, D. Bahnemann, *Chem,. Rev.*, **95** (1995) 69.

[6] J. Yu, L.Qi, *J. Hazardous Mater.*, **169** (2009) 221.

[7] J. Hou, S. Jiao, H. Zhu., *et al., J. Nanopart. Res.* **13** (2011) 5557-5564.

[8] H. F.Cheng, B.B. Huang, Y. Dai, *et al., Langmuir* **26** (2010) 6618- 6624.

[9] H. Xu, L. Zhang *et al., J. Phys. Chem. C* **113** (2009) 1785 - 1790.

[10] X. H. Wang, J.G. Li, H. Kamiyama, Y. Moriyoshi *et al., J. Phys. Chem. B* **110** (2006) 6804 – 6809.

[11] D. Yu, R. Cai, Z. Liu, *Spectrochemica Acta Part A*, **60** (2004) 1617-1622.

[12] S. Xiong, B. Xi, Y. Qian, *J. Phy. Chem. C*, **114** (2010) 14029-14035.

[13] J. Ding, X. Lu, H. Shu, J. Xie, H. Zhang., *Mater. Sci. Eng. B* **171** (2010) 31-34.

[14] J. Zhang, F. Shi, J. Lin, D. Chen, *et al., Chem. Mater.*, **20** (2008) 2937 - 2942.

[15] A. Morrin, O. Ngamna, S.E. Moulton, A.T. Killard, G.G. Wallace, M.R. Smyth. *Electroanalysis,* **17** (2004) 423-430.

[16] O.D. Velev, E.W. Kaler. *Langmuir,* **15** (1999) 3693-3698.

[17] A. Sivanesan, S. Abraham John, *Biosens. Bioelectron,* **23** (2007) 708-713.

[18] D. R.Shankaran, K. Limura, T. Kato, *Sens. Actuators, B* **94** (2003) 73-80.

[19] F. Kong, H. Liu, J. Dong, W. Qian, *Biosens. Bioelectron.* **26** (2011) 1902-1907.

[20] R. J. Mascarenhas, K. V. Reddy, B. E. Kumaraswamy, B. S. Sherigara., *Bull. Electrochem.* **21** (2005) 341-345.

[21] A. Safavi, N. Maleki, O. Moradlou, F. Tajabadi, *Biochem.* **359** (2006) 224-229.

[22] R. S. Babu, *Collids surf.,* B **88** (2011) 755-763.

[23] H. Guangzhi, C. Long, G. Yong, W. Xiaolai, *Electrochem. Acta* **55** (2010) 4711-4716.

[24] S. Shahrokhian, E. Asadian, J. Electroanal. Chem. **636** (2009) 40–46.

[25] M.F. Bergamini, A.L. Santos, N.R. Stradiotto, J. Pharm. Biomed.Anal. **39** (2005) 54–59.

[26] R.B. Kawde, N.B. Laxmeshwar, K.S.V. Santhanam, Sens. Actuators B **23** (1995) 35–39.

[27] J. Mathiyarasu, L. Nyholm, Electroanalysis **22** (2010) 449–454.

[28] V. Hariharan, S. Radhakrishana, C. Sekar, *et al., Talanta* **85** (2011) 2166-2174.

[29] G. Wang, J. Suu, W. Zhang, B. Fang, Microchim Acta 164 (2009) 357-362.

Chapter 6

Summary and Conclusion

The summary of the results and the salient features of the book are given in this chapter. The aim and scope of the book was to prepare different morphologies of $LaFeO_3$ nanostructures by optimizing the various growth parameters both in chemical and physical processes. Also, it was established clear understanding of morphological dependent physical properties such as structural, optical, magnetic and electrical properties. In addition, the different morphologies of $LaFeO_3$ nanostructured powders were analyzed for their technologically significant applications like photocatalysis and biosensing. The goal of the present work has been accomplished successfully and the results are summarized as follows:

Different morphologies of $LaFeO_3$ nanostructures namely, nanocubes, microspheres, nanospheres, flowers, dendrites and nanorods were selectively prepared using surfactants by facile and environment friendly chemical method. A precise control on the morphology of $LaFeO_3$ nanostructures were achieved by using different surfactants and controlling growth parameters. The morphological, structural, compositional and thermal analysis of the different nanostructures formed at various stages and conditions of the reaction was used to propose the possible growth mechanism. $LaFeO_3$ nanostructures such as nanocubes, nanorods and nanospheres have been successfully synthesized through the hydrothermal process without and in the presence of urea and citric acid as structure-directing agents. Systematic studies on the possible growth mechanism were proposed to explain the transformation of nanocubes to nanorods and nanospheres. The formation of $LaFeO_3$ nanostructures mainly went through dissolution–recrystallization process and the self-assembly growth mechanism experiences localized Ostwald ripening process. The structural and morphological changes were confirmed by XRD, Raman and SEM analysis. Similarly, single-crystalline $LaFeO_3$

nanostructures with novel dendrites, flower and microsphere like morphologies have been successfully synthesized through a facile and efficient hydrothermal process. The probable reason for the formation of dendritic structures are the combined effect of diffusion controlled aggregation and oriented attachment of the nanoparticles at favourable sites leading to the formation of such unique morphologies. The orthorhombic structure and flower like morphologies have been confirmed by XRD and electron microscopy analysis respectively. The surfactant controls the solid intermediate phase and further manipulates the growth kinetics. It plays a key role in the formation of flower like nanostructure. Based on the effect of synthesis parameters, the structure and shape of the resultant product and a stepwise growth mechanism of flower like nanostructure was rationally proposed. It includes an initial formation of aggregated nanoparticles via nuclei aggregation. A subsequent site-specific anisotropic growth of nanoplates was occurred at the quasi-equilibrium growth condition dominated by the minimization of surface free energy. The structural and morphological observation of the product shows the formation process of $LaFeO_3$ microspheres mainly went through nucleation, self-assembly, dissolution and reassembly processes. The formation mechanism of $LaFeO_3$ microspheres composed of nanoparticles experiences a localized Ostwald ripening process. All the prepared $LaFeO_3$ nanostructures are thermally stable and does not undergo any structural change even at higher temperature (~1000 °C) and have more adsorbed oxygen or oxygen in the hydroxyl group confirmed from XPS.

After the preparation and systematic characterization, the different morphologies of $LaFeO_3$ nanostructures were studied for their functional properties like surface, optical, magnetic and electrical properties. A comparison of the surface properties of all the morphologies of $LaFeO_3$ nanostructures suggests that shape significantly influences the specific surface area and distribution of pores on the surface of the sample. Among all the morphologies prepared in the present study nanorods were found to have a very high specific surface area. The precursors used in this reaction, the reaction intermediates and their transformation to form the final products was found to be the reasons for the origin of pores and improvement in the surface properties. The optical properties of different morphologies of $LaFeO_3$ nanostructures were studied and it was found that, the absorption cut-off wavelength varied depending on the morphologies. The absorption edge was observed in the range of 580 - 600 nm and the corresponding bandgap energies of $LaFeO_3$ nanostructures such as nanocubes, flowers, dendrites, nanorods, microspheres and nanospheres are in the range of 2.0 - 2.27 eV. It was emphasized the fact that all $LaFeO_3$ nanostructures could serve as a potential visible-light-driven photocatalytic activity. Similarly, studies on the magnetic properties of all the different morphologies of $LaFeO_3$ were made using hysteresis analysis by measuring VSM. The hysteresis feature observed from the VSM measurement confirms that all the $LaFeO_3$ nanostructures prepared in the present study exhibits a weak (or) canted antiferrimagnetism. The important magnetic parameters like saturation magnetization, remanent magnetization and coercivity were determined from the room temperature M-H loop. It was found that all the morphologies of

$LaFeO_3$ shows a significant variation in the coercivity. This could be attributed to the inherent shape anisotropy in these materials. In addition, the variations in magnetic surface anisotropy associated with surface pinning resulted in the increase of coercivity. The observed morphology-dependent magnetic behaviour suggests the magnetic properties of $LaFeO_3$ can be easily tuned for the intended applications. Further, electrical conductivity and dielectric properties of different $LaFeO_3$ nanostructures lacking in the literature has been carried out in the present book. The impedance spectra show two semicircles corresponding to grain and grain boundary conduction and only one semicircle corresponding to the predominant grain boundary contribution. The values of activation energy obtained are characteristic of the hopping of charge carriers. The small increase in the activation energy with the grain size reduction was due to the blocking nature of the grain boundary and lattice imperfections. The anomalous frequency dependence of ε' has been explained on the basis of hopping of both electrons and holes. Based on the physical properties, the prepared nanostructures were evaluated for their efficiency in the photocatalytic degradation and biosensing activity. Since, these applications relay on the surface properties like pore size distribution and specific surface area. The photocatalytic activities of $LaFeO_3$ nanostructures on the degradation of RhB under visible light irradiation found that the nanorods exhibits higher photocatalytic activity compared to TiO_2 and other $LaFeO_3$ morphologies in the visible light region. The enhanced photocatalytic activity of $LaFeO_3$ nanorods was due to its high specific surface area of 84.4 m^2g^{-1} with the pore size distribution of 1-10 nm. The resulting $LaFeO_3$ nanostructures are very promising photocatalysts for degrading organic pollutants under visible light. Most importantly, the obtained $LaFeO_3$ nanostructure with high surface area exhibits high electrocatalytic ability (selectivity, sensitivity, reproducibility and stability) for dopamine reduction when compared to that of other nanomaterials. It was also found the $LaFeO_3$ nanostructures are a good candidate for application as dopamine biosensor in analytical detection.

The present work thus proposes a new dimension on morphology-dependent physical properties and related applications. These intriguing results obtained in the present study describe a clear idea on the formation of different morphologies of $LaFeO_3$ nanostructures and thus provide basic understanding on the methods to control the morphology of these nanostructures. Once the control on the morphology is achieved, morphology-dependent physical properties can be tuned and extended for its functional applications. Finally, we believe that the $LaFeO_3$ nanostructure are financially viable perovskite oxide and it could be a promising material with great potential for applications in many fields.

Index

A

Adsorption – desorption isotherms 70
Amperometric 116
Amperometric determination 121
Atomic force microscopy (AFM) 25

B

BET and BJH methods 25
Biosensing application 99
Biosensing behaviour 12
Bio-sensor applications 4
Brunauer, Emmett and Teller (BET) methods 69

C

Calcination temperature 48, 59
Characterization techniques 17
Chemical preparation technique 18
Chemical synthesis 10
Citric acid concentration 59
Coercive force 81
Compositional analysis 23, 45
Conductivity analysis 92
Controlled synthesis 29
Crystal structure 5, 6
CTAB concentration 51
Cyclic voltammetry 115

D

Dendrite nanostructures 53
Dendritic nanostructures 35
Diamagnetic 7
Dielectric behaviour 95
Differential thermal analysis 24
Dopamine biosensing 116

E

Electrical properties 9, 26, 91
Electroanalytical techniques 114
Electrocatalytic performance 113
Electrocatalytic properties 117
Electrode preparation 117
Electrode surfaces 113

F

Ferromagnetic 8

Fourier transformed infrared spectroscopy 21

G

Growth condition 29

Growth mechanism 48

H

Heterogeneous catalysis process 100

Hydrothermal condition 55

Hydrothermal processing parameters 19

Hydrothermal reaction time 48

Hysteresis loop 81

I

Impedance spectroscopy (IS) 26

L

Linear sweep voltammetry 115

M

Magnetic 5

Magnetic anisotropy 82

Magnetic materials 79

Magnetic properties 7, 26

Magnetocrystalline anisotropy 83

M-H hysterias loop 80

Microspheres 90

Microspheres composite 62

Morphological analysis 24, 31

Morphology dependent 67

Multidomain 84

N

Nanocubes 31

Nanomaterials 1

Nanoplates 37

Nanorods 31

Nanospheres 31, 90

Nanostructured materials 2

Nanostructured perovskite oxide 5

O

Optical properties 5, 9, 26

P

Paramagnetic 7

Perovskite oxide 1

Perovskite oxide photocatalytic materials 101

Photocatalytic 99

Photocatalytic activity 104

Photocatalytic degradation 11

Photodegradation study 99

Physical preparation 11

Physical properties 67

Porosity 68

Precursor 19

Pressure 19

Pseudo-single domain 84

Q

Quantum effect 2

R

Raman spectroscopy 22

Remenant magnetization 81

S

Saturation magnetization 81

Scanning electron microscopy (SEM) 24

Shape anisotropy 83

Single domain 84

Size and shape 3

Specific surface area analysis 25
Structural analysis 21, 42
Superparamagnetic 84
Surface effect 2
Surfactant concentration 55

T

Temperature 19
Temperature dependence 96
Thermal analysis 23, 46
Thermogravimetric analysis 24
Thermogravimetric and differential thermal analysis (TG/DTA) 23
Time 19
Transition electron microscopy (TEM) 24

U

Ultraviolet-diffusion reflectance spectroscopy (UV-DRS) 26

V

Vibrating sample magnetometer (VSM) 26
Visible light photocatalyst 103
Volume ratio 68

X

X-ray diffraction 21
X-ray photoelectron spectroscopy (XPS) 23

9 789359 191300

Printed at Repro India Ltd.